CHRISTO-FICTION

INSURRECTIONS: CRITICAL STUDIES IN RELIGION, POLITICS, AND CULTURE

INSURRECTIONS

CRITICAL STUDIES IN RELIGION, POLITICS, AND CULTURE

Slavoj Žižek, Clayton Crockett, Creston Davis, Jeffrey W. Robbins, Editors

The intersection of religion, politics, and culture is one of the most discussed areas in theory today. It also has the deepest and most wide-ranging impact on the world. Insurrections: Critical Studies in Religion, Politics, and Culture will bring the tools of philosophy and critical theory to the political implications of the religious turn. The series will address a range of religious traditions and political viewpoints in the United States, Europe, and other parts of the world. Without advocating any specific religious or theological stance, the series aims nonetheless to be faithful to the radical emancipatory potential of religion.

For a list of titles in the series see page 275.

François
LARUELLE

CHRISTO-FICTION

THE RUINS OF ATHENS AND JERUSALEM

TRANSLATED BY ROBIN MACKAY

COLUMBIA UNIVERSITY PRESS | NEW YORK

Columbia University Press
Publishers Since 1893
New York   Chichester, West Sussex
cup.columbia.edu
Copyright © 2015 Columbia University Press

Library of Congress Cataloging-in-Publication Data

Laruelle, François.
[Christo-fiction, les ruines d'Athènes et de Jérusalem. English]
Christo-fiction, the ruins of Athens and Jerusalem / François Laruelle; translated
by Robin Mackay.
    pages cm. — (Insurrections: critical studies in religion, politics, and culture)
Includes bibliographical references and index.
ISBN 978-0-231-16724-6 (cloth : alk. paper) — ISBN 978-0-231-53896-1 (e-book)
1. Jesus Christ—Person and offices. 2. Jesus Christ—Gnostic interpretations.
3. Christianity—Philosophy. I. Title.

BT203.L3513 2015
230—dc23

2014031086

Cover image: © Mark Owen / Trevillion Images
Cover and book design: Lisa Hamm

To Robin Mackay and my other British and American translators

CONTENTS

PREFACE
Christianity Stripped Bare by Christ

Our situation may be defined as follows: the war of religions continues, and there will be no Christian return of Christ. If Christianity is the religion of the exit from religions, Christ is the exit from Christianity itself. Let us take a step further, fulfilling this insurrection. It is up to us to invent his impossible coming.

So many christs have been imagined, so many gospels written, plagiarized, copied, canonized or kept secret, unfolded in the light of exegesis or buried in the desert; so many christologies and hermeneutics, so much literature elevated to the dignity of "theological" and sometimes "canonical" texts, that there is nothing excessive in inventing anew his advent, by proposing a christo-fiction that openly declares and advocates itself as such and that consists in a controlled conjugation of renewed scientific-type procedures and old philosophical-type theological models. Why should painters be allowed this right of invention, above and beyond that of mere cataloguing—the right to mix their colors with the flesh and blood of Christ? Why should evangelists, historians, theologians, confessions, churches, and believers have the right to freely fabricate images and stories when there may be a discipline of Christ that, while still fictional, is this time rigorous—a discipline of the faithful grasping themselves as quasi-physicists of the body of Christ, striving to speak his substance and his effects? If Christ cannot be identified with the set of legends that is

Christianity, if he is barely recognizable in his Jewish milieu through his Greek words—a folly for philosophy and a paradox for the Jews—then it is urgent that we revise our categories, which are still those of our beliefs; that we take the leap of thought that is called fidelity, and forge a fiction capable of upholding this fidelity. Above all, not rationalist or materialist disillusionment—that reason is not ours, that materialism is not the materiality of the blood and flesh of Christ. This book, therefore, is not written so as to enrich the treasury of theological knowledge, to verify once more its standards of acceptability, still less those of academic admissibility. Under the punctilious gaze of those standards, its author can only confess shame and confusion.

What is a "rigorous fiction," and is it worth pursuing if it does not essentially reproduce an already recognized truth? Our problem is not that of traditional theology and christology, which, in a sense, we do not seek to challenge; they are first-degree disciplines or symptomal material, like the philosophy with which they are impregnated. Our problem is the *Principle of Sufficient Theology* (PST) that has taken hold of them. Believing themselves sufficient to think Christ-in-person (the Christ of the faithful rather than that of believers), they lack a second dimension— let us say a theologically and christologically nonstandard dimension. It is a question of drawing them out from their lived (*vécu*) of belief and their legendary historical tale, of positing the conditions of their use as means directed toward the raising of a Christ more "authentic" than the legendary Christ because his excess and his effect will be commensurate with "ordinary" or "generic" man—as if the intention were to establish what certain physicists call the "anthropic principle" within the human sphere. A fictional or fictioning fidelity is certainly not a fiction of fidelity. With new means, of nontheological provenance and of what we shall call a "quantum-oriented" order, we have won the right to be atheist religious leaders—that is to say, atheists capable of taking religions from the side where they are usable, and of relating them to that special "subject" called "last instance" that is generic man.

This book belongs to the exoteric genre of theology, but more precisely to a genre of overtly espoused theoretical theo-fiction that can scarcely be said to exist, unless we treat the whole of the Gospels as a literary enterprise of testimony verging on fiction. In relation to Christian theology, christo-fiction distinguishes itself through the removal of belief in its religious and theoretical (that is to say, philosophical) aspects. Theological generality is dissolved in favor of its condensed christic kernel, and above all is reduced to the state of a symptom or raw material for a science of Christ—a science that, despite its positivity, must remain in-Christ. Christo-fiction is distinct from christologies just as the faith of the faithful is distinct from the belief of Christians. More than a literary genre to which the Gospels and all of Christian literature would belong, less than a classical theological genre, christo-fiction is a thought experiment, or more exactly a faith experiment—a practice of nonphilosophical or nonstandard writing that presents and produces its axioms and its rules commensurately with that which they invest. Whence a "demonstration" that is at once globally linear and locally spiraling or resumed.[1]

In what way is this book a faith experiment? Can a fiction be a "testimony" that is "formalized" and thus scientific in spirit? What distinguishes the fiction in the testimonial literature of the Gospels, and ultimately of all Christian literature, from a self-declared christo-fiction based on a theoretical model? In both cases, it is a matter of a truth based upon a positive knowledge.[2] But the paradox lies in an inversion of characteristics: either the truth of testimony is but the continuity of a historical knowledge or the transmission of a spontaneous belief, or it is based upon scientific knowledge, on a model that, albeit contingent in its own way, is rigorous and formalized rather than being a transmitted lore. These are two ways to be contemporary with Christ through the introduction of fiction into truth: either through transmission from a supposed origin (in which case the fiction is imaginary in the literary and psychological sense, even when based on historical facts), or through the controlled production of a cognizance[3] that is nonpositive (in which case fiction is a higher-degree formalization, since it is subtended by scientific knowledges). Neither ancient nor modern, the contemporary of Christ that we seek to be is contemporary either through unfounded, apparent, and hence all the more sufficient belief, or through generic and modelized faith. Two types of truth, two types of subject. Belief is a supposed truth that can be recognized subsequently as being only a fiction. Faith is a truth established a priori under the rigorous conditions of a "well-founded" fiction—that is to say, a fiction founded on the conjugation of a positive knowledge and a theological symptomatology. Perhaps a new kerygma is announced here: belief is the sufficiency of God, but faith is the nonsufficiency of Christ and thus of humans who abase or bring down the sufficiency of God. To sum up all of this, the target and the arrow, we have on one side a Principle of Sufficient Theology, and on the other side (the side from which our struggle is prosecuted) a necessary but nonsufficient faith.

INTRODUCTION
A Gnostic Theology in the Quantum Spirit

What are we to do with the twofold tradition, that of christologies and philosophies, that of the "Christian literature" of the Gospels and of the "lives of Jesus"? If a dialectical reconciliation (Hegel) or a thinking conciliation (Heidegger) have already been tried under the theological authority of philosophy, if exegesis and history have only succeeded in interring and dispersing the message in the inert sands of positivity, what has not yet been done is to conjugate science and theology in a completely new way—in a gnostic spirit that, this time, will make use of quantum theory. How can these two be crossed, as variables equally necessary to thought, stripped of their sufficiency, neither of them any longer claiming to make sense of the other and to dominate it? They are not mere materials, and only abandon their claims within a special matrix for which quantum theory and theology serve only to modelize the whole of the christic message—not to constitute it (it already exists) nor to decide upon its interpretations (there are already too many) but to reorient its use in terms of the humans to whom it is addressed. Thus nothing is lost of the rational rigor necessary for a thought that must refuse all belief, nor of the essence of the kerygma (that is to say, of the faith that defines the faithful). This essay is not a methodological one, a Critique of Pure Theological Reason, nor an exercise in supposedly infallible dogma or discipline, nor indeed one dealing with separate theologemes. It is an essay in the vision-of-Christ-in-Christ and not in the vision-of-the-world-in-the-world. The method, more than ever entangled with its object, treats every separated hermeneutic, axiomatic, and dogmatic as properties of the same object become complex through the integration of its dimensions. However it is certainly not a return to our old enemy, the dialectic, arisen from the empty tomb of philosophy. If you must have a governing thesis or a principle then here it is, in all its brutality: *the fusion of christology and quantum physics "under" quantum theory in its generic power, and no longer under theology.* This is called a matrix, and serves to determine, albeit as indeterminate or underdetermined, the cognizance (our faith) of that X that we call Christ-in-person, on the basis of the material of his images, fables, or sayings, his conceptual christophanies.

New tasks are prescribed by this placing of Christ's message into the gnostic matrix:

- To treat the specular doublet or theo-christo-logical difference as the essence of Christianity. The set of its effects is theoretically, but also practically, concentrated in the Principle of Sufficient Theology that governs Christianity.

- To suspend the supposed validity of the two basic disciplines when left to their positivity and its spontaneous claims. Christ is neither theologian nor philosopher, nor even the founder of a new "human science." But it was necessary to demonstrate this, to furnish ourselves with a suitable, independently validated method, and to draw all the consequences of it, not just state it confusedly or pay lip service to it.

- To manifest Christ's message in a nonhermeneutical manner as being neither Greek nor Jewish, as being unintelligible by way of Logos and Torah separately or mixed together, but intelligible if Logos and Torah are canonically conjugated as variables—that is to say, as properties posited for Christ but which are only determinable when considered outside of any melange, with one excluding the other. To theology's vicious manner of thinking, we oppose the inverse products of these interpretations, and give a concrete translation of them, with both topological and phenomenological aspects, in terms of relations between transcendence and immanence, Law and Freedom, proximity and infinite distance.

■ To demonstrate how Christ is the founder of a science of humans that is nonphilosophical and nontheological, but generic—that is to say, man-oriented or underdetermined by what we call "being-in-last-humanity" or generic man as "last instance."

■ To manifest the generic essence of Christ, which is to render humans idempotent, so that each may verify, through Christ, his equality to another. Christ is the internal law of composition that makes it possible for a human to equalize himself to another and to lead him from rebellion or revolution to heresy. We call Christ-in-person the mediate-without-mediation that transforms the believer into a "faithful-in-Christ" and humanity into a "mystical," that is to say, nonphilosophical, body.

■ To reinterpret the scene of the Cross and the Resurrection. The refusal of history and its "vulgar" temporality necessitates the establishment of a new causal order for these events: It is the Resurrection that underdetermines, as prior-to-first, the primary order of the Crucifixion and the Entombment. The Resurrection is essentially a *surrection* or an ascending that must no longer be understood as a philosophical transcending or an elevation in exteriority, already viciously impregnated with theology and opening the way to every dialectic.

■ The erected Cross is the inverted objective image of the Resurrection, which is the real movement that has been falsified by the theology and the belief that uphold it, in an apparent or objective movement. The apparent sacrifice of Christ is really the sacrifice of God, the end of the rivalry between Father and Son, the birth of Equals.

■ To reinterpret the distinction between belief and faith. Theo-christological belief is the specular doublet of Christian representation, and moves within Greco-Judaic melanges of transcendence. Immanent faith is the generic—that is to say, human or idempotent in-the-last-instance—phenomenon that distinguishes the faithful from believers.

■ To rename this whole set of axioms as being those of a gnostic but nonreligious usage of Christ, of his kerygma and of his faithful. The gnostic spirit is the fusion of science and philosophy, of knowledge and salvation, in equal shares; the operation of the subtraction of the faithful from the world, the sacrifice of the "Evil God," and the glorification of Christ as mediate-without-mediation for the faithful.

Renewed, non-Christian gnosis proceeds via the substitution of contemporary knowledge (quantum theory, as model of thought) for the purely conceptual science of theology. In this regard one might call it, strictly speaking, a quantum deconstruction and a generic transformation of theology.

CHRISTO-FICTION

One

A GENERIC REPETITION OF GNOSIS

To Desuture Christ from Theology

Our basic equation is as follows: *Christ = science of Christ = gnosis*, along with its corollary, *gnosis vs. theology and christology*. That Christ is simply the name of the science of Christ, that its other name is gnosis, and that "gnostic theology" therefore means that theology is abased (without being completely negated) as object of gnosis—nothing in these radical axioms belongs to any known Christianity. Christ is the name that tears gnosis from christology, leaving the latter to float in its indeterminate theological ground. We place all theology globally under the last instance of Christ.

Gnosis—that ancient name, still loaded with religious phantasms, abhorred by the "Old Believers" of the West; a name we take up once again, inscribing it entirely under the name of "Christ," in so doing voiding it of certain aspects of its doctrinal and theological content from which, in its early historical forms, it did not manage to liberate itself. We understand it in a nontheological and even nonreligious sense, as the substitution, in another place and with other functions, of Christ for God. The dominant theological plan of salvation, the work of God and the basis of "Christianity," borrows its essential definitions and philosophical foundations from Greek ontology. In this respect christology is a concentrate of those Greek philosophical prejudices from which we must tear away its kernel—a kernel that, for lack of a better term, we shall call "christic."

The principle is to desuture Christ from theology; we do not recognize christology understood as the domain of trinitarian theology, but only as a science-in-Christ that will replace such contexts, of which it is (in a sense that will become clear later) the real content. A science-in-Christ, however, cannot ignore theology, but instead assigns it an entirely new role. The Platonic reinvention of philosophy after early physics and then that of theology through its transition via Aristotle exalt it and sterilize it twice over: by placing it in a specular confrontation with mathematics, and by inserting it into a great "modern" planification of the orders of knowledge, orders it will be charged with dominating. This enthusiastic exaltation and this theological sterilization of philosophy are, for gnosis, only materials, material to be transformed in view of a more adequate knowledge and defense of humans. Doubtless, religious and theological phenomena precede humans, who are thrown into them; but then, humans precede the science of these phenomena. Better armed with rigor and humanity, radical (not absolute) gnosis is the real content (which does not at all mean to say the historical or textual content) of the science of humans as taught by Christ. Christology is therefore only a particular, indifferent material in this science-in-Christ, which is also a science of Christ. For this science, God, in any case, is no longer anything more than an object governed by the rules of a certain objectivity, what we may call a quasi-quantum-theoretical objectivity—God is both a variable and an "occasional" motif of this science. Atheism is, in all respects, a hasty, mediocre, and thoughtless solution, passive and naive, as is materialism itself. To put it in the most paradoxical way possible, gnosis as we conceive it is a new "Reformation" guided by contemporary science; it is relieved of what remained of transcendent religiosity in the protestation against it; it converts this mythology into philo-fiction or christo-fiction. It cuts more or less vigorously between Christ and God, dismembers the Trinity, undoes that too-tightly-knotted packet of Three Persons. Theology is the concern of the Father and of his substitutes, and goes by way of monarchy; gnosis is the concern of the Son, and thus of humans in their equality, and goes by way of the democracy of generic brothers. Its problem is the defense of humans, not just the question of what they themselves can make of the rest of the world into which they are thrown. Rather than a modern thought, it is a contemporary thought—one better suited, for example, to new forms of faith, possibly including evangelical ones, but more assuredly to the form of a thought that crosses science with philosophy, otherwise than evangelically.

GNOSIS, THOSE WHO VANQUISHED IT, AND THOSE WHO WERE VANQUISHED

To discuss a subject such as Christ, his death, and his resurrection, without any theological preparation, as we advocate doing, bracketing out the theological point of view as the dominant point of view but one that has no deciding role here—all of this evidently requires a good deal of naivety, and even unconscious humor. All that is demanded of us is a gnostic preparation that must place us in the correct stance, that of inventive fidelity. In this "epistemic" framework, gnosis represents the bringing together of positive knowledges, including philosophy, within a human-oriented cognizance of them.

How to conserve what is essential of the fundamental human and scientific ends of gnosis, at least reduced to their main invariant factors? For it is obvious that the introduction of relatively recent scientific means, unknown to ancient gnosis, such as quantum principles and nonphilosophy, will end up inflecting these ends themselves almost as much as the means of their implementation. Of humanity, of its defense, of the use it can make of contemporary sciences, we still know nothing solid—we have only contradictory philosophical and religious projections, in a certain sense only *variables* or *properties* of humanity. The wager of this renewal is founded precisely upon the person and the message of Christ as inventor of a science of humans (but not "the human sciences"). It is true that it is a tradition of early Christianity, and of certain currents of Christianity today, to conceive Christ as the founder of a new conception of man. It cannot, however, be a question of redoing, varying, or amending what has been tried a thousand times already, insofar as it was done at the whim of philosophies and religious systems. New means, such as quantum physics and nonstandard philosophy, authorize us to suspend these early attempts, insofar as they are impregnated with the ancient, metaphysical concept of science, and with the religious concept of Christ—at least to suspend their sufficiency, if not their materiality. This bracketing-out of the spirit of the ancient knowledge of Christ, and in particular that of the whole theological tradition, will seem scandalous and rather glib after the infinite debates of theologians and believers. Let it be understood that we do not intend to declare that everyone has been mistaken about the significance of Christ—in a sense, all of that must serve as materials for us. We propose another hypothesis—that of a Christ-science or a science "in-Christ" that appeals in principle to theology, at least as symptomal material. The only principled solution is to take up once more the problem of Christ as a scientific and human—that is to say, gnostic and generic—problem.

Gnosis is known for having been vanquished by the Church as dogma and as institution; it has about it that mystery of lost, ancient doctrines that lie at rest under the dogmatic marble, reduced to the aura they give off. Its theology has so often been condemned as heretical, its great thinkers forgotten or reviled, sometimes burned to death, that it is hard to imagine what a gnostic theology might be. Thus a work of updating, not only of rehabilitation, would have to be undertaken in order to prepare the notion of a more contemporary gnosis. Now, this work of exhumation began long ago—it is the achievement of those historians who dedicated themselves to the origins of Christianity. But it is a question of knowing whether it is a corpse that has been exhumed, to be delivered to the forensic pathologists of religion, or whether gnosis is "resurrected" and can "rise" into the heaven of theory as a new possibility. It has remained confined to specialists, to the historians whose highly erudite affair it was; and, on the public side, the impoverished understanding of it as a sect-like esoteric doctrine has never really been corrected. The old division of its reception between scholarly gnosis and popular, nefarious gnosis has been continually refined, but never overcome. Such is the usual destiny of the "mysteries" that religions cultivate in their practice and their theology, the difference being that the fate of gnosis seems to be sealed, despite the scholarly interest that has taken hold of it. In a certain way, it is impossible to crack open a doctrine that has always been opposed, and always for "good reasons"—reasons that in a sense remain pertinent, since they are the reasons of force, of theoretical domination and of dogmatism, of condemnation and of scorn, of institutionalization and of the norm. It is not a question of rehabilitating gnosis by reversing classic theological irreversibility, attempting a coup that would end up adopting the mistakes of its adversary. On the contrary, we must conserve the secret of gnosis, but at the same time find some way to make intelligible its unintelligibility and its unlearned character, we must discover the means to conserve and manifest its secret without destroying it qua secret with an inadequate, rationalist light. All of this is a part of what we hope to achieve theoretically: precisely a gnostic-type knowledge.

This project immediately involves quantum-theoretical means, which must be adopted in order to be able to say that we are dealing with a science. It is a known principle that, in quantum-theoretical terms, to clarify a supposedly given or existent secret is automatically to undetermine it in and through this very knowledge. But in fact we must be less abrupt and dogmatic: it is possible to clarify a secret in a quantum-theoretical manner without absolutely destroying it (this would, rather, be the effect of philosophy) if one can establish the quantum "law" of that phenomenon—that is to say, what we shall call its "state vector." And this is what we try to do with and in the science of Christ: to establish the state vector of Christ on the basis of his data or the data of his words in the "Logos" mode and the "Torah" mode. So it is not a question of reconciling, through some new mediation, scientists and simple ones[1] (the quantum formalism opposes this) but of finding *the fusion point of simple ones and theory, where the masses of the gnostic faithful seize theology—and not only the theology of simple souls.* The Churches detest this type of project, which does away with mediations and makes for "heretics"—a good sign for us, a pertinent criterion that allows us to identify the true adversaries.

Neither is it a matter of attempting a historical resurrection conforming to the great systems that illustrated ancient gnosis through a commentary on its "evangelists." Our incompetence as historians is both confirmed and voluntary. The ancient gnostic systems, as grandiose as they may have been in comparison to our philosophical systems, are, like the latter, marked by Greek problematics that we intend, if not to do away with, at least to make only a controlled and limited use of. More generally it is a matter of elaborating a theology as a science that is directed, as a function of the gnostic orientation, toward the person of Christ and not toward God—without it being for all that a mere question of christology as domain of theology. Our problem is that of a theology placed under the christic condition of the "last instance," a science of Christ radically distinct from theology, capable of using theological means even as it distinguishes itself formally from them. This science of Christ is the means of "humiliating" theology, of abasing it, of making it the servant of Christ and of humans. It is not (for example) the project of a "cultural" reeducation of philosophy by sending the masses to the school of mathematics. To reeducate theology does not necessarily entail abasing it, or making it a mere means for the science of Christ—it could just as well mean the affirmation of a haughty and solitary mastery. Inversely, to use theology while no longer making of it a discourse that glorifies the name of God is not necessarily to negate it; it is just to deprive it of its claims to omnipotence, as concentrated in the *Principle of Sufficient Theology* or of *Theological Sufficiency*. The conservation of the term "theology" in the expression "gnostic theology" is a little surprising, but hardly more than it is in Denys the Areopagite's "mystical theology." The formula is perhaps more direct than the equivalent terms "nontheology" and "nonstandard theology"; it has the virtue of making a gesture of acknowledgment to the vanquished of thought whose memory we intend to reawaken in "modernized" or, more exactly, "contemporary" forms. A gnostic theology of Christ, articulating the science of Christ and the decline of traditional theology, signifies the decline of theocentrism in favor of what can no longer be a christocentrism that leaves undisturbed the structure as a whole. We try to "reactivate" or resume (*rélancer*) the inspiration of gnosis using contemporary means, which, as we shall see, must therefore be (in accordance with this inspiration) scientific rather than religious, generic rather than philosophical.

Let us define the broad characteristics of a gnostic ethics without religious or metaphysical foundation, but with only a generic foundation. (1) It sutures ethics to a complex cognizance that cannot be defined by the dominant Logos or Reason, nor by the transcendence of the Other, and still less by man as biological animal. Ethics is not an effect or a predicate of reason or of life, but the operation of what humans-in-body can do; ethics belongs to the ultimate substance of humans. (2) If there is an ethics, then there must be evil, unhappiness, or malaise, evildoing, evil-thinking, evil-acting, evil-dwelling (for example)—this is the paradox of an ethics for the philosophers that we all are, virtually. It is not a question of their conversation or their logical argumentation, of their wisdom or heroism; it is an immanent therapeutics of man in the world by man qua open to the universe—not an autonomous therapeutics like that of philosophy, limiting, purifying, and improving itself. Its cause has its seat neither on Earth nor on High. (3) Unlike philosophy, which is attached to a dominant discipline and therefore to all disciplines without exception, it does not derive from a monodisciplinary knowledge, from biology or politics; and its apparently decisive relation to quantum physics is subject to special, very restrictive conditions. Generic ethics crosses local knowledges two by two, forms relations between disciplines (always with philosophy) in a state of collision; or, in any case, it affirms their nonseparability. It is an ethics and a politics of interdisciplines, of the most opposed heterogeneous knowledges that humans have acquired. (4) It is generic because it is oriented toward the a priori defense of all humans in all situations or all possible universes (not only in the world and outside the world), following the criminal experiences of the twentieth century, which gave ample proof of philosophy's revisionism. (5) It renounces religious metaphysics, and the theology of the Good and of God. It is a gnostic trait to introduce into man a necessary ingredient of the non-Good, but a relative one—a trait that is perceived by the Church and the State, united together, as an evil—an evil in principle, even a willful heretical derangement of the natural philosophico-Christian order. It renounces the PST and finds in Christ the paradigm of the victim in the state of in-surrection, of the Sacrificed who is also the Resurrected; it extends the sacrifice to God and to sufficient theology, in favor of the Resurrected; it makes of Christ freed from God the nonsufficient or nontheological real, the necessary but nonsufficient cause of salvation. Gnostic-atheist salvation is, precisely, now assured by Christ-without-God alone; and this is a necessary but nonsufficient condition of salvation. Christ is this nonsufficient, nondivine condition of salvation—we need the cooperation of human subjects implicated in the operation at least as agents or actors. But our context is no longer that of the Christian faith, stuck in between the prescience of an omniscient God and the intervention of human decision. The quantum correlation or unique "state vector" is that of the underdetermining human condition of salvation and of the occasion of human will.

THE CONTEMPORARY ORIENTATION OF GNOSIS

In order to establish that "gnostic theology" must be scientific, we do not refer directly to the letter of gnostic systems—they are not our affair, we are not concerned with the historical—but to their broadest governing motifs, which appear immediately with the thesis of the generic-oriented character of the science of Christ. It is a question of reformulating the salvific character of gnostic knowledge using contemporary theoretical means, and of removing gnosis from its religious and even its Christian context—and of doing so by recentering it on Christ.

We call gnostic, in a broad or formalized sense, a thought that presents the following characteristic traits, whose enumeration should aid us in "modernizing" gnosis and removing its religious definition. (1) It defines man through the unique relation to a certain knowledge that has a complex status—both learned and unlearned, a knowledge that is his immanent substance—rather than through a more or less transcendent predicate such as rational, metaphysical, political animal, and so on. That kind of association philosophically divides generic humanity, its simple animality being definitively rejected as the obscure ground of life, while it is consigned, on the other hand, to the divine transcendence of the Logos—this is a doublet, not a duality by superposition. (2) This human knowledge is called generic because it conjugates, without syncretism but by means of a complex matrix, the most strongly opposed heterogeneous knowledges, those that form the framework of all disciplines: science and philosophy, which furnish it with its twofold model. In reality we treat gnosis as a necessary ingredient of the relative non-Good that belongs to the Good, and which is necessary in order to deploy the latter as generic.

THE DEPLOYMENT OF GNOSIS: FROM POSITIVE KNOWLEDGES TO COGNIZANCE

We distinguish two states or two uses of any knowledge or epistemic material: an encyclopedic and sufficient use, which also includes philosophy; and a salvific use, which alone is a generic knowledge. Historical gnosis is by definition a doctrine that is difficult to identify, like philosophy itself, admitting of multiple variants, certain of which are of a scientific type. But it is a mixture with a unitary aim, known for its struggle against the ecclesio-centered Christianity of the tradition; it is clearly not a duality affecting the use or the destiny of knowledges (even empirical knowledge) or all epistemic material. Our so-called nonphilosophical or nonstandard use respects several of its typical tendencies but seeks to renew them, or even to found rigorously the distinction between their uses. In order to do so it will, paradoxically, submit this epistemic material to a science that is itself drawn from the contemporary encyclopedia but that will be used in a more complex, less encyclopedic manner than its initiators intended.

Generally speaking, gnosis is an unlearned or untaught knowledge that is supposed to inhabit the deepest heart of humans and to assure their salvation. But this definition is too totalizing, unitary, and religious, and disregards the theoretically complex conditions of salvation and its relations to the sciences. The concept of gnosis must be differentiated and displaced from its Greek basis, by way of the contemporary sciences. Its new basis or its primary stratum is thus the set of contemporary knowledges, simply transmitted, received, and practiced within their proper order, which is that of the world or of philosophy, a collective knowledge acquired by humanity or by individual workers subject to the laws of capital and its incessant accumulation. In this broader sense, encyclopedic gnosis contains a multiplicity of natural knowledges that, sedimented on the ground or the history of humanity, become "unlearned" or quasinatural. Obviously they have been acquired, taught, and transmitted, but in the continuous form of a tradition, a sedimentation that belongs to global or world humanity, and thus also to the individual—but not at all to "subjects," properly (generically) speaking.

This knowledge, which although primary is not especially or solely mathematical, in general has no use or finality outside of the world and its philosophical, theological form—which does not take us beyond the encyclopedic. But it can be taken into account instead as a simple means or productive force, stripped of its traditional finalities and its sufficiency, in view of the constitution of an entirely other knowing—a "cognizance" that cannot be called either philosophical or theological, but only generic, because it is related to the human genus, whose responsibility is to make the best use of it, a use that this time no longer concerns teaching and transmission, but instead the invention of the safeguarding of those beings living upon the earth. The form that governs its extraction and its consumption is thus no longer that of philosophy, but that of the human genus that is to be preserved. Knowledge gives rise to a cognizance of salvation—which is the reason why gnosis, which defends humans and, on this basis, living beings in general, is not at all "anthropocentric" in the usual sense. It is a cognizance of the last instance, the ultimatum in the face of which the destruction of human life may possibly not come to pass.

Of these two uses of the same knowledges, the first might be called actual in terms of production but virtual in usage—it is produced actually in history, but no longer serves the salvation of humans. The second is virtual—it is a cognizance that one *is*, no doubt, but that one must also have, or acquire. Renewed in a context that is more scientific than religious, more generic than philosophical, gnosis displaces the old division of knowledge between being and having, apparently inverting the old order while affirming that they are the same thing, since the knowledge one has of generic humanity is also a knowledge that one already *is*, but is yet to acquire cognizance of. It is a knowledge that one does not yet know as a cognizance of salvation; we shall oppose it to philosophy's formula, which speaks of a "knowing that one does not know." This formula really signifies a certain sufficiency, signifying that it is better to know "nothing" or to know the nothing of knowing than not to know at all. Philosophy makes of knowledge an absolute imperative that contains the knowledge of nonknowledge itself, as nonknowledge. Gnosis cuts down this absolute will to knowledge, and radicalizes or differentiates between knowledge and the cognizance of this knowledge.

EPISTEMIC REPETITION: CHANGING THE SCALE OF KNOWLEDGES

Gnosis is generally understood as a unitary and solid, almost Parmenidean body of knowledge, in which Platonic hierarchies and distinctions are laminated. But this is to forget the excesses of its imagination, the multiplicity of its mythological entities, of the sects through which it was disseminated, and the great theoretical systems to which it gave birth. It is above all to misunderstand its epistemic essence, which is that of a duality condensed in the formula "knowledge-of-salvation," a soteriological cognizance that takes on its full meaning when confronted with the maxim of philosophical or worldly wisdom formulated thus: "take into care beings as a whole." The more philosophy dedicates itself to ontology, turns around Being, and alienates itself in that which is, the more gnosis dedicates itself to man and to science, and turns around the axis that it forms with these two things, and that is called Christ.

We thus intend to change the paradigm of thought, admitting that our theological knowledge no longer has the force of legitimation; that, as regards the Christ we are concerned with, only his most simple sayings any longer have any force for us, the force of the philosophically unheard-of; that we have to invent, as rigorously and faithfully as possible, our Christ. At best we have a model in the Gospels, but no example. The error of Church theology is precisely to have made of the evangelical model, which could well have been a model in the scientific rather than the Platonic and philosophical sense, an example to imitate, in a stance of rivalry—as if Christ were not so much to be created as imitated. What is demanded of us is precisely a fidelity to his sayings in their simplicity, and the effort of inventing a thought that "goes with"—*with* Christ. This simplicity of Christ's sayings is not necessarily defined by that of his listeners. It stems from the symbolic force which we shall say is that of the "superposition" of their expression and their meaning, a simplicity one finds in the resumed outbursts of certain mystics, or in "confessions of faith." Our knowledge of Christ exists in a symptomal form that does not distinguish between the letter and the overentangled meaning of the apostles' texts. For this nonseparability or nonlocalization to be recognized, the natural or quasi-physical ground of human being must be manifested, transformed into a generic (not anthropological or philosophical) cognizance. As a minimum we shall disregard, at least provisionally, Saint Paul's commentaries and his planetary considerations, on the borders of the Judaism from which he issues and the universal Christianity that he tries to found.

The sense of this repetition of the fundamental features of gnosis is destined to be transformed into an "epistemic" experimentation in salvation that is (at least) performative: a practical and lived theology. Gnosis is not just a discourse to be conserved in the memory or in the museum of Christianity. We proceed with its repetition in the first place so as to verify our theoretical means, those that produce gnostic knowledge, and to prepare what will elsewhere be called an "operatory field" or "experimental vessel." The theoretical matrix used here has a twofold implication: we are implicated in it as subject in such a way that it itself is implicated and transformed by our experiment. This twofold implication must however not remain in a vague state but must "turn generic." Gnosis is the duality of a primary or positive knowledge implicated as a means for human transformation or self-cognizance. We must prepare ourselves for a cognizance of *we-the-humans*—but not so that this knowledge may exceed itself once more, gain power, and modelize itself, as happens in philosophy. On the contrary, it will have to subtract itself from its spontaneous reflection, to abase itself or be abased, removed from its self-cognizance as practiced by philosophy. For this is what it is about, against all dogmatisms and vain promises: *evaluating the chances, the "probability," of an epistemic salvation of humans.*

To sum up, we usually distinguish knowledge that one *is* from knowledge that one *has*. But there are two possible scales upon which this distinction may be made, one philosophically centered, the other nonphilosophically centered. From the latter point of view we consider the knowledge that is acquired by the positive sciences, and that forms the excess of our everyday context, as having become knowledge of the infused or ontological world; it loses its acquired and laborious character—even the most demanding experimental science is thus "frozen" and sedimented in a quasi-materiality that one *is*. This is the *elementary and new* level of our concept of gnosis, which does not overlap with the old one since it transcends and homogenizes the traditional divisions, even those of doxa, of positive science, and of the Idea, which are all parts of it.

What is more, there is the radical concept of a gnosis that is of the order of cognizance, perhaps of truth, rather than of knowledge. It is also acquired through an operation on elementary knowledge, treated this time as material or symptom, and it also becomes, in a sense, infused. But if one has or acquires the primary knowledge in the world so as to become it or be it, one has or one acquires cognizance as outside of the world, on the basis of this primary knowledge, and by treating it as a quantum-formalized raw material. It is a question, in this second phase, of gnosis as veritative knowledge of the salvation of humans such as it is determined in-the-last-instance by them and no longer by philosophy as form of the world. It also is infused, but indirectly so, and possesses even less immediacy than the first phase of gnosis. This is to expose acquired knowledges in the context of the world and of its validity for a certain human practice—to submit them not to the eternity of a divine gaze as Platonic philosophers do, but to a generic futurality. This is the meaning of gnosis, this exposure to messianity but not to God—an atheist, nonreligious, and non-Platonic gnosis, an act of the salvation of knowledges that tears them from their native milieu whence they draw their sufficiency.

Their generic usage is not their absolute self-exposure to a dialectic, but their radical reprise in terms of a futurality commensurate with humans.

SUFFICIENT NONKNOWLEDGE AND NONSUFFICIENT KNOWLEDGE: FROM DIALECTICS TO QUANTUM THOUGHT

If gnosis is a knowledge that ensures salvation, what exactly are its objects and its means? Ancient gnosis holds that the response is self-evident since it is variously religious—Jewish, Platonic, or Persian. For we who have at our disposal other theoretical means more complex than those of the Greeks, their conjugation with a theology immediately penetrated by syncretic religious influences is of hardly any possible use—it is no more than a mysterious memory or an object for historians. How to "repeat" it, submit it to a reprise through procedures of knowledge that have gained in extension and in precision but are still capable of being centered on the person of Christ as it demands and still in view of a salvation through some cognizance? Gnosis operates on a duality condensed into the notion of a knowledge-that-saves. We have shown that it is a double knowledge, a duality, and not a unique knowledge, as it might seem in the overdetermined religious context that secretly plays the role of a second term, functioning as an unperceived material, this religious context that goes without saying. Remember that a fully deployed gnosis rests (1) on a ground of validated but contingent knowledge, (2) which provides the "natural" ground of humanity and assures it of what is, in a certain way, its primary substance (3) which does not know itself or which is not directly interested in man, (4) which must therefore be put to work for and by man as a means for his salvation. Obviously, both ancient and modern philosophical thought know of dualities that seem to be of the same type, to be deployed according to the same logic, until one examines them more closely: for example, the idea of knowing that one does not know, or that one knows nothing, the duality of knowledge and thought, faith and knowledge, manual work and intellectual work, intuition and concept, practice and theory, esoteric and exoteric, metaphysics and thinking thought. But finally, with this unitary concept, it will be a matter of thinking that which we know we do not know. To think or know that one does not know, to know that one does not know or knows nothing, one must nevertheless know implicitly that one does not know, at least via a third party like Socrates or a philosopher like Plato, before replacing this absolute nothingness with confirmed knowledge. The philosophical axiom underlying all of these solutions consists in postulating the implicit power of nonknowing as already being knowledge in the Other or in the Self, the postulation of the capacity of nonknowing to be deployed in the form of a higher-grade knowing. Let us call this claim the *Principle of Sufficient Nonknowledge*—it is the very heart of philosophy.

It is not enough to assert nonknowledge and to raise it into a standard—rather, a nonsufficient knowledge than a sufficient nonknowledge. Gnosis inverts the problem of philosophy, or formulates it more honestly. Why suppose, through bad faith and hypocrisy, that we do not know or that we know nothing, when manifestly we know far too many things, all of which burden the world into which we are thrown? Rather know for certain that one knows nothing or very little than claim not to know at all. More straightforwardly, let us admit that we have real primary knowledge, which includes that nonknowledge of which philosophers speak, and admit that the difference or the duality is always between two forms of knowing. This is not a contradiction, like Being and Nothingness, but a unilateral complementarity. If there must be a difference to produce, it will be that of the knowledge that we have as beings (thus, that knowledge that, in a sense, we *are*, but without having it for ourselves). The generic difference lies between that knowledge whose human destination remains unknown to us and a cognizance of that knowledge that we will be, or, far more exactly, that we will be by *means* of that "primary" knowledge.

So what does it mean to say, in the order of cognizance, "not to know this primary knowledge," not to be cognizant of it? The subject does not need to have read Marx to know what it costs him not to have the *means of existence*, to have an "existence without means"—for example, theoretical contemplation with no means of invention. It is rather a matter of reducing this knowledge, even if it is "pure" like the matheme, to a means or a procedure stripped of its supposedly sufficient proper finalities. It is not sufficient to reflect, philosophically, this knowledge that one is without having it. Gnosis short-circuits transcendental or absolute reflection, which still belongs in every way to primary knowledge. Without reflecting it once more and potentiating it further, to infinity, it forces it to change destination, to bifurcate from its spontaneous finality which is philosophical idealism, to submit itself generically to humans rather than to a God. At this point we cannot yet understand how gnosis forces primary or positive knowledge to enter into a procedure that is not at all reflexive or in specular torsion, but "vectoriell (*vectoriale*),"[2] or how generic ex-sistence is not a trajectory in the void but an emergence at its root—a radical, not absolute, insurrection. Far from being a corpuscular knowledge like a spark or macroscopic like a pearl as the religious gnostics imagined, contenting themselves with doubling it with a specular and empty transcendence and therefore remaining within the unitary thematic of the philosophical flash, generic existence's insurrectionary character subtracts it from such positivity. It inverts the religious schema. It is in essence primarily a vector rather than a thing or an object, but one that brings with it the equivalent of a microscopic or partial (we say "quartial") object to which positive knowledge is reduced. Deployed gnosis is thus indeed double, but precisely without forming a doublet—it is instead something like a superposition or a complementarity. It is the duality of a primary ontico-ontological knowledge, given more or less positively by the different disciplines, and a veritative use of this knowledge in terms of man, a duality stemming from quantum properties such as superposition and, as we shall see, noncommutativity.

We shall call positive any knowledge that one does not yet know how to orient generically as a means of salvation, a knowledge learned or taught through a transcendent tradition that is materialized as immanent or in a becoming-unlearned. As to the cognizance of this knowledge, it is "radical," not at all absolute like philosophy, but in a certain way "untaught," just practiced. To say it rigorously through the nonphilosophical procedure, it is "nonlearned." That which has not been taught to us yet which we know as in a mirror in Christianity must now be practiced by other means, quantum means, in this way conserving its secret status.

THE CONJUGATION OF SCIENCE AND THEOLOGY

To make a generic science of ancient gnosis, three sources must be combined: philosophy and its theological modality as object or material; science in the form of quantum theory, no longer as their object but, on the contrary, to underdetermine them and take them in turn as object; and finally man as participant and stakeholder ("last instance") in this combination, as interested in it. All of these general characteristics will see their own sphere of existence and action modified by their generic assemblage. Let us detail these conditions. What is necessary is:

1. That science in the form of so-called objective, axiomatic, or deduced knowledge should intersect with philosophy, cutting it straightway and "orthogonally" from itself. Not in two more or less equal halves, but unilaterally, science fusing with a part of philosophy (the lived) that it carries with it, whereas the other part of philosophy (consisting in the structures that make it an ideal body) is at once indirectly dependent on this fusion and independent of it as appearance of sufficiency or of the in-itself.

2. That this lived science should thus be subtractive or underdetermining of philosophy in itself qua spontaneous philosophy of science, upon which it would wish to impose its categories and its determinations, in particular that of the "all" and those which derive from it. Philosophy is a doublet of itself, a dominant double transcendence; it is to be abased or reduced to a simple transcendence.

3. That its principal act ("superposition") should be constitutive of a scientific subjectivity designed to replace in an entirely other site (that of the "last instance") and with entirely other functions (those of underdetermination) philosophical, egological, and cogitative subjectivity.

4. That this science should be capable, through its quantum orientation, not of determining philosophy reciprocally or simply inverting it, but of underdetermining it, subtracting it from (over)determination and thus rendering "philosophically" indeterminate the objects it treats of (including the subject). These objects continue to exist; they are not negated, but are simplified or reduced to a transcendence without doublet, which is true "vectoriell" immanence (insurrection).

5. That man should have a fundamental interest in this science and its intersection with philosophy, in objectivity as much as in the lived; that, as science fusing with the lived, it appropriates not all of the individual or philosophical subject, but a part of it, and that it becomes a subject-science (*science-sujet*). This is the Feuerbach and Marx side of things: man is not just a part of nature, but makes nature human.

Ultimately, science as generic gnosis is the fusion of theology and quantum theory, a fusion quantum-theoretically (that is to say, generically) underdetermined.

THE GENERIC ORIENTATION OF GNOSIS

This complex device is capable of irretrievably tearing humans away from philosophy alone, precisely because it takes account of philosophy as one of their variables or properties. The generic, thanks to its quantum aspect, is noncommutable with the philosophical, although it does make use of it. Philosophy's flippancy with regard to man is unfathomable—it oscillates between a narrowness of perspective, a reductive cosmic prejudice, and a stupid self-assurance pro and contra that caricature called "humanism." It has never been made for man, but always for the world in which it incarcerates him, the being into which it throws him, the nature within which it inscribes him, the unconscious to which it subjects him or through which it shreds him up as "subject," the society in which it dissolves him, the mathematics of which it makes him a "function"— here is the most profound alienation of humans, the one that governs all the others. Facile philosophies of alienation took the work only halfway; philosophical alienation would have had to have been excavated all the way down to generic man. Certain disciplines have really taken man into their care without necessarily placing him at the center of their theory— philosophers such as Hobbes, Rousseau, and Marx, and primarily science fiction and gnosis, certain currents of the theology of the Eastern Church. Apart from that, all is world, chaos, substance, physis, eternal return, spirit, cogito and cogitat, consciousness, subjectum, truth. It is no longer a matter of contesting science, or even objectivity, which is not necessarily to say positivity, an access by right to the human lived; it is always philosophy or a philosophical position that takes up the means of science so as to submit man, and not only the individual, to it. Man is not commutable with philosophy, which is but one of his means.

As to the historical and religious gnosis of the Ancients, it must be corrected. It is often defined in terms of the importing of Greek and Platonic concepts into Christianity, the duality body/ideality or body/soul replacing the new Christian duality of flesh/spirit, far more positive and broader since the flesh includes the body and ideality in a positive order, to which it opposes that of spirit. We know that it is a deviation from "authentic" Christianity, which also affected the "world" with which the problem of the body had interfered. Here also the positive relation of heaven and the world created by God, a world that is but the contingent place and cause (and sometimes the sign) of sin but that is subject to redemption, was "hijacked" by gnosis (and not only the Manichean kind) into an irrevocable condemnation of the world and of bodies. For reasons that are, precisely, scientific (albeit physical rather than unilaterally mathematical), and for reasons of a finer description of bodies by phenomenologies—in other words, for reasons that are globally more "Aristotelian" than "Platonic"—this deviation of gnosis in relation to Christianity no longer makes sense for us, and we must deliver the old gnosis itself from these terms.

What we call generic gnosis is a reconfiguration of the old kind, against the Platonic deviation that in turn became "Christianity"—a return to Christ without Christianity. It has the twofold and yet quite unique task of being at once non-Christian and non-Platonic. Of conjugating the Platonic detestation of the world and the necessity of humans' being forced to love it sufficiently to live in it and to transform it without being able to leave it or flee from it (this is what remains of Christianity), but equally without having faith in it (this is what remains of religious gnosis), according to it only what is called, precisely, "belief." This conjugation takes place within a generic matrix that, to say it differently once more, multiplies the variables of Christianity and gnosis by each other, fuses them by underdetermining their unity through gnosis. It goes without saying that more than ever, and faithfully to the spirit of gnosis, our task is to invent the Christ who will be our contemporary. Must we add that, despite many barely discernable ambiguities, we do not entirely conflate free gnostic invention and sectarian servitude? Gnosis is a theory and a practice of the invention of life, which has until now been practiced in a religious context. It was itself invented for reasons of religion, combined with philosophy and Judaism; but it is necessary, now that many of our means have changed, to found it more rigorously as nonreligious gnosis. To found it insofar as it itself is not capable of founding a philosophy.

THE PHYSICS OF CHRIST AND THE PRINCIPLE OF NONSUFFICIENT SCIENCE

What science could respond to these demands and assure us of its pertinence for an object such as Christ and the phenomena in which he is given, if not a purely conceptual "science" like theology—that is to say, a false science? The sayings of Christ, the Crucifixion, the Resurrection, the Ascension, faith, grace, the body, life, and so on—these phenomena are as concrete (physical, even) as one could hope for; there is nothing philosophical here—that is to say, nothing separated into opposites and then reconciled. All is truly "material-and-spiritual," and precisely to rid us of philosophical positions, all is "*materiel* (*matériel*)"[3] through superposition, not through dialectic. These phenomena are made for the "Simple" because they themselves are "simple," even if they are destined to be grasped by a nonsufficient understanding, an understanding stripped of the Principles of Philosophical and Mathematical Sufficiency (PPS and PMS). There is a materiel phenomenality of Christ. His formulations and the vicissitudes of his history belong to a quantum "materiality" of a new type that escapes both the simplicity of analysis and the complexity of philosophical and transcendent syntheses. In any case, all that counts is the a priori under which the science is going to apprehend its object. To measure the science of Christ against the empirical appearances of history or against a Galilean epistemology, as did the classics (Hobbes, Spinoza), is now a fruitless procedure in relation to our instruments. To define him through philosophical positions such as materialism or idealism is almost as pointless, and comes down to simply entering into a vicious or hermeneutic circle. Now, a physics of christic phenomena could not resolve itself into a diversely nuanced philosophical hermeneutics, even if we say sometimes that this physics, in its own interpretation of the real, can be aided by and can use theology as an "input" and as a hermeneutic variable. This implies not only that it is this science alone that determines its object, but that it is necessarily a special science, a physics that is not positive—not exactly conceptual, but delivered from its positive, sufficient, or spontaneous use. For the great problem, what we shall call a nonepistemological problem, is that of a *nonsufficient science*. Just as there is a PPS, there is a PMS. A generic science must be capable of implicating its theoretical apparatus into its object rather than separating them. The science of Christ is nonseparable from Christ, it cannot objectivate him in just any manner, but remains a physics with a minimum of mathematical productive force, and does not become a hermeneutics even when it makes use of the theological discourses that have formed around Christ. It is possible to include theological materials in a physics if the latter has as its object a materiality of words and inseparably ideal (and) material events. It is obviously a question of that special physics that is quantum physics. We isolate the principles of a nonsufficient quantum thought, the rational or principial kernel of physics, and invest it in the christic materiality of some of the events and utterances of the Christ-being. Classical theology, like conceptual science in the philosophical manner, must be and can be replaced by a physics of bodies or of christic corporeality, and can borrow new procedures of more rigorous experimentation on these phenomena than the traditional and autoscopic hermeneutics of faith by faith. Gnostic theology, in short, is the rigorous knowledge of the christic phenomenon through a combinatorial of theology and physics, but one that is underdetermined by Christ as messianity. The messianity in which Christ is resolved is that always-complete never-closed opening called faith or fidelity. There was in gnosis a messianity that was lost or drowned under the accumulation of dogmas and mythological images, and that it is possible to resume or reactivate.

QUANTUM PREPARATION

Traditional theology seeks to be a conceptual science. It immediately finds itself caught up in the distinctions between the conceptual and the intuitive, and in their confusions and mélanges, as illustrated philosophically by Leibniz. These latter are pursued in Kant despite his efforts to make a distinction between intuition and concept. For Kant tries at once to maintain and to overcome this distinction in the transcendental imagination and the schematism, but will never have *really* overcome it by way of the latter, which represents what we could call an amphiboly of the transcendental itself, albeit one superior to those of reflection. Classical theology thrives on this type of amphiboly, mediated by the transcendental or by the Hegelian concept as self-mediation. It is of the greatest importance to perceive the complexity of these philosophical solutions, whether Kantian-schematizing or Hegelian-mediatizing. Even nineteenth-century philosophical christologies, which drove on further into this Hegelian and more broadly speaking philosophical terrain, did not surpass this type of transcendent complexity. The thesis that we oppose to this type of solution is as follows: the various melanges of intuition and concept, of given and intelligibility, of empiricism and rationalism (Kant)—whether analytic or synthetic, it matters little here—all belong to the corpuscular or macroscopic style, to take up these old indicative terms of physics; they are neither wave nor particle. The classic duality between intuition and concept, illustrated by the memorable essays of Leibniz and Kant, and schematism and self-mediation along with them fall entirely under the corpuscular model of reality, which is but one side of the latter, its Newtonian side. A physics of Christ, if it does not wish to risk giving rise to a physicalism, can no longer treat the wave/particle complementarity like a schematism that, as can easily be appreciated, is a macroscopic duality with a certain affinity with the religious and philosophical context of Christianity, its sensible/spiritual dualities and its psychological imaginary. God and Christ, the sayings and the events of the Apostles, the dogmas to which they give rise—all of this material must be treated in the form of dualities or complementarities of a quantum type, or more precisely a vectoriell or unilateral type. The christic science that replaces christology at once changes the face of theology as science. Insofar as it is built on principles drawn from the quantum-theoretical model, it is constructed on the one hand from algebraic properties such as idempotence and the imaginary number represented geometrically by vectors, and it ceases to be logico-formal and gives rise to an algebraic formalism, without PMS; and on the other hand, it is constructed on a matter of the lived, which is no longer given intuitively, but is given materielly by wave and by particle, this lived being the substance of christic phenomena. It would be dangerous to invert this materiel formalism into a formal materialism that conjugated a materialist position and a mathematico-logical model, and was inscribed entirely within philosophy.

We must, then, extract from positive quantum physics, with the aid of the philosophical variable (the two being conjugated in a generic matrix) the kernel of "quantum thought," not with computer software (*logiciel*) but with "quantware" (*quantiel*), a vectoriellity for new thoughts. Quantum thought does not mean that science can "think" (an absurd formula), but that it becomes a *means* for thought, once we make use of philosophy for this extraction—an operation that is the inverse of the typical procedure of philosophy, with science cutting the corpuscle of philosophy, slicing through the All, and demanding that both be superposed so that philosophy is able to not become "science" (another absurdity symmetrical with the first one), but become a *means* for the science of Christ, or enter into its service. It is in this way that science and philosophy enter into a common labor as means of a nonstandard thought. In other words, one cannot exit from the amphiboly of traditional theology, which mixes science and philosophy to the benefit of philosophy, abusively idealizing or indeed materializing Christ, except through their conjugation as variables of the object = X named "Christ."

It is true that if one stops at this stage, with the inverse products of science and philosophy, one has indeed entered into a matrixial and noncommutative conception of theology, but not yet a generic conception. Here the conjugation is not a mediatization or a schematization, but it must be submitted to a special condition that is the "reprise" (rather than the repetition) of quantum thought that will henceforth find itself prior-to-priority, as before-the-forefront through this operation of reprise, but that nonetheless remains inseparable from the philosophy that it makes use of and that makes use of it, each as means for the other. One might have been able to think, up until now, that these multiplications were mere "partial identifications" between quantum physics and theology, but it is no longer a matter of this once the quantum variable itself is resumed. For this resumption, far from being a repetition, difference, or identification (all of which are philosophical operators), is a "superposition" in the algebraic and quantum sense of the term. A superposition produced ontologically from the idempotent, algebraic, and nonmetaphysical One, which is capable of supporting an addition to itself while remaining "itself" or forming a vectoriell immanence, without becoming a doublet. The sterile addition of a synthesis that passes via an analysis without sinking into it, of an analysis that passes via a synthesis without stopping at it. It is valid only as vectoriell immanence, not as philosophical transcendence. It makes the One and only the One-in-One with some Being or with some Other. The One that is but One, addable or superposable with itself, is underdetermined in relation to its doubling and its metaphysical identifications as Being and Other as overdetermining instances in the face of the One that underdetermines them. It is the decline or the generic abasement of theological transcendence, its underpowering that is idempotence in relation to omnipotence.

INCARNATION: SCHEMATIZATION OR CLONING?

God "made" man—doesn't this great axiom borrow its real content from an operation of the cloning of Christ and the faithful that takes place in the matrix? The real or phenomenal content of the macroscopic schematization of God in man by Christ is what we might call a cloning, realized by an immanent act, acting on and in a material of transcendent origin like the lived of the subject or of philosophical belief, and reducing its doublet to simplicity. The Christian version of the schematization of God in human nature is the bad transcendent fusion of terms, a macroscopic fusion with a spectacular and repetitive result. Whether it is the hyper-macroscopic God or whether it is the apparently more modest and immanent transcendental imagination, these operators still borrow from the schema of Christ, or from the human, worldly, and psychological predicates in which they alienate themselves, the better to save the humanity of these predicates. In this operation, Christ remains God, or rather becomes the God who was and remains God—he rejoins a Father who is not really alienated but subsists as he is. The sinful transcendence of man has doubtless been humiliated and abased, but not that of God, who has only apparently been abased; there has been no real "alienation" of God, or "objectivation" of him as Marx would say. This is a macroscopic game that changes almost nothing. The result risks being tautological, or at best giving rise to a reinforcement of power: that of God, and that of the transcendental imagination—that little God hidden in the depths of the soul, that operator secluded in the shadows with which it enshrouds itself. The Moderns have their mythology, for sure more "rational" than that of the gnostics: they have boosted power and domination; to the traditional quadripartite of causes they have added that of a new demiurge, the transcendental imagination, the anthropological mirror and doublet of God.

Incarnation is a dialectical concept beyond all human comprehension, which is why this miracle fascinates the multitudes who do not understand it. For it to become intelligible, as faith and fidelity are, it must be reduced to a physical order, and physics already has enough paradoxes of its own, which it is able to recognize as such, without needing to add to them the supplement of belief. The "generic matrix" is that device, sufficiently "miraculous" in its order, in which vectorial (*vectorielle*) algebra turns into or rather is "made" into generic or human vectoriellity (*vectorialité*), a fusion with a materiellity that it informs without first having to resolve the problem of its validity for matter. It is only if the matter of the phenomenon (as Kant would say) is given empirically that the problem of the schematism of the concept poses itself—that is to say, only within the corpuscular or macroscopic framework of rationalism. However, in the generic matrix (and this is where it differs from the transcendental aesthetic), vectoriellity acts as a subtraction or abasement of transcendence (of the complete or doubled circle of philosophy) *through the imaginary or complex number or through the quarter-turn*. In general it is easier to subtract a predicate from matter than to numerically add a form to it, addition and superposition being a subtraction from both matter and philosophical form.

We must therefore read the effectiveness of the matrix in the complex of Crucifixion, Resurrection, and Ascension not in a historical manner—that is to say, according to an operation of superposition hardly visible in the Ascension—but as that which retroactively clones the lived of the human subject. No analytic or synthetic construction of the matrix—that is, of the complex of the Cross in all its dimensions. The Cross thus understood is a machine that is already running before our interpretation takes place; and it is this machine that produces the clone of the generic subject or the fusion, as Christ, of contraries, of the vector as the simple transcending of the human lived. The cloning is the work of the generic matrix (as immanent operation of the production of Christ or of the Faithful as mediate-without-mediation) or of the underdetermined fusion of the variables that are the Forces of Production and Relations of Production submitted to the concreteness of the matrix. Cloning is a physical concept of Incarnation, and cannot be explained by way of philosophical or theological operations of divine transcendence.

God "made" man: this enigma of enigmas is too mysterious not to have been interpreted by the shortest routes—both those of myth and of those aspects of myth that are imported into theology, and those of science, certain of whose statements, rigorous as they may be, remain in the neighborhood of myth and are capable of reducing it without, for all that, tipping into a "Voltairean" reduction. The generic matrix furnishes a general principle of the intelligibility of Christ-thought, a principle we shall find again in regard to those other great foundational mysteries that are the Cross and the Resurrection, alongside the Incarnation. The at once quantum and generic explanation necessitates a special methodological precaution as to the time of explanation in relation to the phenomena to be explained. An "explanation" that follows the course of the time or of the history recounted by these mysteries will surely be just as mysterious as its objects. It will be a fantastic story, like theology, which traces itself from the stories of the Gospels and sublimates them. A science does not have the right to mime its object or to specularly reproduce it give or take certain explanations, variations, or conceptual commentaries, even if they be speculative doublets, which are perfectly sterile beliefs. "Demystification" must be radical to also be, very simply, a demystification of the philosophical substance of theology. This is truer yet for a science of the quantum order than for the predominantly anthropological and determinist sciences of the age of Enlightenment and of what subsists of the "philosophical" in Heideggerian phenomenology and right up to Derridean deconstruction. To tear out the last roots of temporal and historical determinism that support theology, we must agree not to judge belief by belief, but faith by faith, or more exactly by the last-instance messianity of Christ. We shall utilize the "imaginary" of the Apostles, qua historical story and spontaneous foundation of faith, in another space proper to science, specifically the science of Christ, which, we should recall, is not a purely historical experience, one of empirical testimony documented by disciples who will "scholarize" the "teaching" of Christ by spreading it throughout the world. And even more so, it is not that transcendental history to which German christologies have habituated us. This space is that of an experience of Christ; it is an experimental, that is to say, performative, space that implies our own faith, and that calls for relatively precise procedures. One of the first precautions to take is to note that the temporality of the real of Christ and the Apostles is not deterministic, otherwise it would become an abyss of theological aporias—that is to say, precisely an abyss of dialectical mysteries. To uproot the imaginary of the Apostles and extract its kernel of generic human reality, it is necessary not simply to invert the course of the history that is recounted (that would hardly gain us anything) but to think this history (that is to say, the proven knowledge we may have of it) as underdetermined precisely by Christ, organized and ordered by futurality or messianity. There will be no hermeneutic circle of faith of a philosophical or believing subject that ultimately founds itself, but only the objective appearance of a circle of faith determined in-the-last-instance by Christ's messianity. Thus the Resurrection and the Ascension will be the phenomena whose futurality justifies or explains the Cross, doing away with its aspect of barbarous sacrifice.

Understood as superposition and as mediate-without-mediation, not as identification or schematization, Christ escapes the mastery of theology and the concupiscence of priests who always have the macroscopic resources to recrucify him by seeking to penetrate his so clearly evident secret. *Christ under-goes as his body-on-the-cross, visible to all and yet secret; the Cross is idempotence as potency of the Same*; it is the superposition of the horizontality of Logos and the verticality of Judaism, their orthogonality—more entangled than a knot, for a knot can always be untied, at least locally.

CHRISTIC INSURRECTION AND THE ABASEMENT OF GOD

What is the most general effect of the gnostic reprise, and not just of the philosophical repetition of gnosis, of its texts and its history, a hermeneutics that does not concern us here? It is a non-Christian deplanification of the history of salvation, an insurrection in the very principles. Not a simple inversion of the relation of Father and Son, as dreamed of by certain millenarian movements within the Christian religion, it is a matter of the promotion of the Son to the status of the prior-to-first cause or "last instance" of a new history which would finally be that of humans and consequently that of the abasement of the sufficiency of God, placed in turn under condition of Christ. This insurrection is prosecuted against the Principle of Sufficient God, or the Principle of Sufficient Theology. The Christian Good News does nothing more than reensure the priority (an almost natural priority) of Christ as son and as sacrificial substitute for sinful humanity. The outcome testifies, as (much later) will the revolutionary outcome of Marxism, to its failure in regard to its theoretical conditions, in the absence of a new global problematic capable of succeeding Greco-Judaic thought by taking stock of it. It is almost impossible for a "cultural revolution" to succeed if it is not accompanied—preceded, perhaps—by a real theoretical "insurrection" or a new, adequate framework of thought. All we can say is that Christ is not responsible for the disaster that followed him, and that he succeeded exactly insofar as Christianity foundered in its worldly ossification. There is no revolution "in" history as long as history itself is not "revolutionized" or revolted against. Like Kierkegaard, we think that Christ is the "absolute fact." However, we are content to say "radical fact," which should put history back on a human footing by proposing another intelligibility for it, obviously not an empirical one, but one of second degree. It is a matter, first, of identifying, otherwise than by following the thread of historical determinism or of the "life of Jesus," the quasi-scientific significance of his message, and of discovering within it a new paradigm of faith, of action, and of the stance of the faithful when they are in a generic body and engaged with the world. The Acts of the Cross, above all that of Resurrection, have no spiritual sense except a materiel or lived one, and the faithful must imitate them in an adequate way. In Christ as in the behavior of the faithful, we must elucidate the immanent phenomenon of the Resurrection not by way of the philosophical model of repetition or revolution, but as a revolt or insurrection whose immediate effect is the bringing down or "decline" of God. The phenomenal content of the Resurrection is a generic reprise, an immanent leap, rather than the old transcendent leap; and its correlate is the fall of the God of the Old Testament into human generic immanence. From the macroscopic and omnipotent that he believed himself to be (monotheist mythology), he becomes (or, better still, undergoes as) particulate, as quartial God, reduced to a negative quarter of his old omniscience and power. To put God under determining condition is not to negate God purely and simply (a gesture of thought just as hasty as the simple refusal of theology and philosophy); it is to give oneself a chance to raise theology to what it has always claimed to be: a science of God that treats God as the generic object it is. Christ is our revelation as faithful humans, and God the revelation of the correlative object of this fidelity. The abasement of God and the revolt of Christ or of messianity correspond to the fall of philosophy or theology. The gnostic repetition finds its content and purport in this human insurrection and in the correlative abasement of the old God. Such is the meaning of a properly gnostic "atheism," the refusal of the "sufficient" goodness of God.

BEYOND ATHENS AND JERUSALEM

We defend the concept of a gnosis broadened beyond its historical and theoretical delimitation, a gnosis we might call a gnosis of the Last-Humanity. There are two "evil gods" and not just one: the ancient Jewish God whose exorbitant pretensions justify his having featured in the first rank of our enemies. There is also a Greek logos-God facing him. Greek paganism is not the absence of religion, just the absence of absolute monotheism; it is the affirmation of a multiple God, an affirmation that we reject through the broadened gnostic refusal set out by what we might call christic or generic gnosis. That Logos is a God and a Law was recognized much later and in a temporality other than that of the Torah, but it contributes nonetheless to constituting Christianity as a heritage. To consider the Eternal Return of the Same, the culmination of a unique and pagan God as multiple, as a mere concept or an ancient mythical oddity of Nietzsche's, as he brings together the essence of Western philosophy as "theology," seems to us to smack of an academic superficiality and an irresponsibility that barely counter the Church's claims of a Greek source or reason. We shall have to interpret Christ as a generic matrix that gives the Cross its true meaning, at least so long as the latter is understood as an effect of the Resurrection, rather than the Resurrection being understood as an effect or consequence of the Cross. If the Cross relates to the theo-christo-logical doublet and signifies God's omnipotence, the victory of Christ on the Cross signifies that he is of another nature, that he must be understood otherwise, through categories at once more rigorous (scientific, even) and more generic—for example, as a collider for those two rationalities that he makes interfere and vibrate, producing messianity and faith. They have been interpreted by Christianity in a way we might well call primitive and "noncivilized," through a scene of crucifixion, of sacrifice, which can only lead us to expect the downfall of their omnipotence and that of the Cross itself. This gnosis is of a christic interpretation: it is no longer Christ who will be a mere object of historical gnosis.

The concept of gnosis must itself also be reworked in the direction of a quantum and generic amplitude.

STRUGGLE AND FIDELITY

We are fighting on two fronts: against Socratic theoreticism and onto-logical wisdom, and against Judaic ethics and the Law. These founda-tional religious legacies, the Real as Idea or as Law, do not have the same meaning now, as if one could choose one or the other indifferently. But the struggle against Christianity is another thing altogether, a little dif-ferent and more complex, and with it is initiated the steepest decline of transcendence into mediation, a sort of regression to immanence (and one that will have political or "gnostic" effects). The ideal and the possibility of an immanent "life" no doubt emerged within a religious context, but Christianity is perhaps the sole religion that can negate itself, immanen-tize itself, interiorize itself to the point of denying its divine transcen-dence via incarnation as the death of the transcendent god, as sacrifice of the ancient religious ground in favor of the Christ-subject delivered to solitude and abandonment. From this radical immanentization that "returns" as positive undergoing, we can (like certain contemporary his-torians of science) draw out the possibility of a modern, non-Greek science, but more broadly, the possibility of a non-Judaism and a non-Christianity—in sum, a nonphilosophy. We have seen, however, without drawing any other consequence than its Hegelian and dialectical (that is to say, Judaeo-Greek) reappropriation, that with this new paradigm of sci-entific immanence Christianity demanded a redefinition of thought itself in its very foundations. It became possible to reformulate it according to a causality neither Greek nor Jewish, to transform its usage of science and of religion, mélanges of which it would no longer tolerate as a style of thought. The problem of "philosophizing in Christ" is not resolved so long as to philosophize is not "in-Christ" but remains within the prior-ity of the Idea. Incarnation is the model or (more precisely) modeliza-tion of a true prior-to-priority of the Real so immanent that it separates itself from the All. Christianity and Judaism are only relatively opposed to philosophy—one through an immanence without any true means, the other through an excess of transcendence. The in-Christ supposes the dissolution of religious pagan mixtures that are revived in philosophical christologies. The new governing formula is to philosophize, to Judaize, and perhaps to "mathematize" (if this formula is understood correctly) in-Christ as in-One. The true formula, that which will have filled a space or a void abusively hidden and filled in by philosophy, is thus to "underphilosophize in-Christ." We do not understand the "in-" as transcendent incarnation but as a lived-(of)-immanence that is the real of incarnation. It is not a matter of a historical Christ or of a Christ idealized by religion or Platonized by philosophy, but of Christ as Stranger-subject or Son of Man-in-person. To draw something new from Christianity itself, and to do so "in-Christ," we must understand the story of the Gospels as a modelization of a radically immanent Christ, and must double Christianity in a generic christo-fiction and a christo-centrism that will be its religious modelization.

Christ offers the chance of a defection from the positivity of religion, but also from that of the philosopher, because there is indeed one, even if Paul diverted the christic message from its meaning, rejudaized it too quickly without truly Judaizing-in-Christ. Christ announced the end not of Judaism but of all religion, and perhaps of monotheism, for a generic monohumanism. Not even a Protestant reduction of religion in favor of Scripture, for this is a last remainder of fetishism and of the quest for consensus. Christ is not the critic of religions, he is their consummation as immanent and lived, who leaves them in the state of residues—that is to say, symptoms and models with secondary functions. He is not even the deconstructor of philosophies, since he calls into question the ultimate presuppositions of deconstruction. This is why the great problem now plays out, at best, between christo-centrism and christo-fiction, the latter not being a negation of the historical Christ but his devaluation so that he is relevant exclusively as a modelization. The passage from the interiority of Incarnation to the immanent cloning of the Son of Man allows for the positing of a new Real for philosophy and for Judaism. Radical incarnation obliges the separation of the Real as prior-to-priority, more than ex-sistant (Lacan) to the Gospels, from "his" thought. The messianic lived neither thinks nor speaks; it is the world or philosophy that think, and hence the Christ-subject, whose essence is no longer the Idea, the Law, the Other, Scripture.

Two

THE IDEA OF A SCIENCE-IN-CHRIST
Christ, Science, and Their Gnostic Suture

To "philosophize in Christ" is one of those injunctions of which philosophy boasts so many—but this time a Lutheran (and Pauline) one against Plato and the Greeks. It means recognizing the grandeur of reason, since one must philosophize but at the same time place reason in conflict with faith. This would also be the thesis (like so many we find in Marxism) of a nonphilosophical practice of philosophy (in this case a Christian one). Understood in this way, this maxim is susceptible to innumerable equivocations so long as the rigorous concept of Christ is not itself established independently of philosophy. We understand this maxim of a nonphilosophical practice of philosophy as the obligation to establish a science, which nonetheless would not at all be neutral, of Christianity—a science that engages with the Christian affair but not like a hermeneutics does, and not only as a positive science of it.

To philosophize in Christ may be nothing but a still-theological slogan, and this indeed is how it has been understood. But in that case it is not fundamentally new, and merely recenters on Christ a new, vagabond, speculative, and theocentric theology. We understand this formula otherwise, as a call to transform from top to bottom in an experimental or current practice (somewhat like the mystics did) what we call "theology," to make it into a lived and faithful affair. Rather than a theology accompanied by a minimum of faith or a faith accompanied by a minimum of theology, always more or less easy to externally suture, we understand theology as a lived experience in the contemporary sense of experience—precisely as experimental, abandoning the norms of philosophical validity but not all philosophy. Can theology be a life rather than the contemplation of a possible life? A work according-to-faith rather than a faith without work or a work without faith? Can faith be an immanent praxis of theology and thus something like a theology implemented or put to work by faith? Such an inflection (more than simply "lived": formalized) of theology as a work operative in each of the faithful signifies a taking leave of the Church as apparatus of mediation, and of theology as apparatus of the Church.

For faith to be the condition of the immanence of the most theoretical works, the praxis faithful to theology, it is necessary to dismember, in a manner itself nontheological, the theo-christo-logical doublet that structures it and makes for its authority and sufficiency. If this dismemberment itself is no longer to obey a theological practice impregnated by philosophy, it must be carried out in an entirely other manner. The formula "to philosophize in Christ" resonates like the affect of a distance impossible to bridge but that must be bridged, the call of a blank space that is radical in a certain sense, of a lack into which precisely the Church and theology, who have a horror of the void, have been cast, as mediations designed to resuture the whole formula. But another operation was possible: this blank space testifies, no doubt, to a void, to an implied term, but there is no reason to reform a whole according to the same mode of unification. The term that is lacking, because it is a stranger to philosophy and to the theology of Christ, is that of "science." Yet another term testifies to it, precisely yet illegibly, in theology and its transcendent ontological presuppositions (that of the "in": "in-Christ," which indicates an immanence of Christ and an insertion) to philosophize and thus theologize in this immanence. What we shall call the generic (or christic) matrix is the apparatus, a scientific and experimental apparatus (physical, and more precisely quantum-physical) that succeeds in the dismembering of the theo-christo-logical doublet and the inclusion of theology itself in this immanence that bears the name of Christ. This formula therefore calls for a change of theoretical element for theology. A program which is that of the fusion of Christ and quantum science in a generic thought.

Does "science-in-Christ" thus testify more crucially to the grandeur of Reason, or to an effort to cut down that grandeur? We resolve this aporia, which radicalizes the injunction to "philosophize in Christ" so that it pits Luther, Pascal, Plato, and many others against one another, by introducing between philosophy and Christ this third term which was not expected in such a form, that of science. And by identifying in Christ and in him alone the subject bearing a *generic science*—that is to say, a subjective science, a science of religions. Such is our primary thesis. To philosophize in Christ does not affect the grandeur of Reason if it really is a question of a science-in-Christ. This would be to place philosophy, and therefore theology, as complexes of knowledge and belief, under condition of a new theoretical but practical stance, to determine them and transform them by way of a science whose blinding yet invisible, clear yet silent principles would be contributed by Christ. Christ is not just a religious model to be imitated in his existence or in his sufferings, the founder of a new religion that ceaselessly returns to interpret and solicit him, but the author of logia that must be read as the protocols and axioms of a new science of humans—and humans, moreover, insofar as they are committed, as beings of beliefs and rites, to the world. It is to this Christ that we submit religious thought, as one submits an object to the principles of science. An important nuance here: it is not the Christian religion, still less "Christian science," that is the science of other religions; it is Christ who announces the protocols of a science for all religions, Christianity included. Christianity is here no more than what we could call a "formal" or else "primary" religion, to be placed under condition, a christic condition. Needless to say, this is not a positive science, although it has identifiable principles and procedures taken from a contemporary science that is both experimental and formal.

Why speak of a "science of Christianity" and of other religions? Has faith ever contributed a science, or indeed allowed itself to be the object of a science? Historical reasons, although quite superficial, can nevertheless put us on the right track. The Christ-event is none other than that of the emergence of faith, against the Greco-Pagan and Judaic beliefs that it alone is capable of transforming so as to place them in the generic service of humans. And this struggle against beliefs in the name of faith maintains the closest of links with modern science—that is to say, physics. If philosophy and mathematics since Plato have been conjoined like twins in a mirror, faith and mathematical physics have done so otherwise, and together signal the entry into the Cartesian modernity of subjective certainty (Heidegger) for which "Platonism" is no longer anything but a means in the service of thought, rather than an inspiration. The theoretical (in the broadest sense) conjuncture having changed a great deal, this association now has different means at its disposal, and can propose new objectives for itself. One will not be surprised therefore to see that here we require certain algebraic (neither logical nor philosophical) properties used in quantum physics, which represent the hard kernel of contemporary physics, and which we hope are suitable for producing the rigorous understanding of faith. Conjugated with the Lutheran imperative, this new perspective can only signify a taking leave of philosophy as exclusive or theoreticist contemplation of faith. Should we say, then, that this is Christianity in the science of Christianity? Not exactly, and it is this that distinguishes it from a hermeneutics: there is Christ-in-person in the science of Christianity and of other religions. "In-person" does not mean the individual, but a generic universality various echoes of which still inhabit the Trinity of the "three persons," despite its Greek mask. If prior-to-priority is granted to Christ and not to Christianity in the science of Christianity, the circle is broken, but only on condition that we can suture science and Christ. Now, this suture, if it is the horizon of modernity, is more profoundly a gnostic affair. We retain from the gnostics, in the hope of escaping from the hell into which the Church and Philosophy discarded them, two fundamental axioms. When formulated in a manner suitable to our conjuncture, these axioms will suffice to sketch out the most general framework of this essay: (1) A certain special suture of science and the subject—a fusion, why not, of theory and the faithful masses, defines man as generic knowledge rather than as the man of "Greco-Christian humanism." (2) This knowledge is faith itself, in subjects delivered to the world, and is the practical source of their salvation.

To invoke physical "science" in these domains seems a rather pointless positivist provocation, or a return to the old ruses of metaphysics as "science" as in German idealism, or at best "as rigorous science." But "the sciences" cannot be reduced to mathematics or to hard (or even soft) sciences, positive sciences in general. We understand the sciences otherwise than taken up in the forking (the Caudine Forks, even) of the paradigm positive/transcendental. Quantum physics displaces this paradigm by imposing a nontranscendental immanence on the one hand, and on the other by implying that the subjectivity of the observer must necessarily be taken into account in the preparation of our faith-experiment. It is the possibility of "subjects-sciences (*sciences-sujets*)" that has an antidogmatic and generic pertinence distinct from philosophy (despite the transcendental interpretations of quantum physics)—and distinct also from positive sciences, which do not include the subject in their procedures but only in their objects, and which furnish knowledge, not truth or the conditions for truth. To do this, at the same time as we transform theology, we must transform the positive usage of (for example) quantum physics. We must recognize an affinity between certain scientific principles to be specified and the logia of Christ. Obviously we are not, above all, going to say that Christ is an individual subject in the traditional matter, the subject of a science or of a philosophy—all of that is out of the question for the generic subject-science.

To philosophize "in" Christ? The solution already depends on what we understand by this "in": through, according to, because of, for? We are evidently involved in a theory of immanence that Christian philosophers controlled through the subjective interiority of faith, others through that of the transcendental ego, and yet others through the interiority of the mystical body of the Church or that of scriptural texts. Whence idealist christologies, which are mere interpretations of Scriptures using the means of philosophy, or the dogmatisms of the Church. For us it will instead be a matter (after a great deal of explanation) of philosophizing "in" that science which *is* the Christ-event. Since Christ is identified here with a generic science confronting theology, there will be always and only two terms and not numerically three—what we have introduced as a third apparent term will not be a mediation or a philosophical synthesis. We do not define a christological position, but a scientific stance of Christ that takes as its object all religions and in particular their two "substantial" poles, paganism as *illustrated* by philosophy in its Greek origins or by the Logos, and monotheism as *illustrated* by Judaism and the Torah—and their more "formal" Christian synthesis, the Christianity that is but the symptom of Christ-effects.

Let us say straightaway (and this will be a second thesis) that if the science-according-to-Christ is a device with only two terms, then Christ is or contributes a type of intelligence that is faith or messianity itself. If faith or messianity constitute the understanding and the critique of religions, it is up to us to find the axioms of faith, the principles of messianity.

Our objective is to formulate laws that, without being those of representation, even religious representation, are capable of explaining the latter as that which falls to humans. For on the side of its object, this science is that of humans, of course, but humans qua subjects involved in the world rather than with Being or even *being*-in-the-world. If Christ can be credited with a science, it is that of the world and of its last object—this is what distinguishes him from the ontology that occupies itself with the median zone of beings and Being. All the same, for the humans that we are, it is indeed religions, along with their philosophies and their theologies, that are the form of expression of the world which Christ gave us to comprehend. Whence a third, yet more polemical thesis deduced from the principles: Christianity is the set of standardizations of the Church and of the appearances of Dogma—at the limit, the whole set of religious travesties of the message and the "person" of Christ interpreted in the mirror of the world.

It will be said of Christ, to sum up his actions, that he is the faithful or the messiah "in-the-last-instance," *the last messiah as before-first*, an expression that means less than ever a cause or event supposedly hidden behind Christianity, and still less a transcendental foundation. The idea is rather to treat him at once as a constant of the scientific type, and thus as "objective" in the sense of his being an invariant for all possible human science, and also as the fulfilling of subjective functions. The principle of this science is however not some synthesis, a Hegelian synthesis for example, of subject and object; it is a *quantum of faith*, constant by definition (which does not necessarily mean quantitative), and of the lived (which is not to say of any lived whatsoever or of worldly belief). An objective constant fabricated within the lived, a lived of faith objectively informed. In this way, faith or messianity qua constant preserves the grandeur of Reason, which is not conflated with that of philosophy (whose sufficiency is, for its part, weakened). The antinomy of the concept of science-in-Christ or, strictly speaking, of "generic theology" is thus resolved as follows: the dephilosophized grandeur of Reason remains to science, with which it canonically conjugates, so that here at least it behaves like a subject or a lived that *is* faith, whereas the sufficiency to be abased remains to philosophy and theology. The sufficiency of philosophy and the grandeur of Reason have too often been conflated. We address another conjugation of them, which we call "generic" and no longer "philosophical" in an exclusive sense: *the fusion of science and philosophy under science rather than under philosophy or theology.*

THEOLOGICALLY UNINTELLIGIBLE FAITH AND ITS INTELLIGIBILITY

How can we speak of a non-Christian science with "Christianity" as its object, and other religions through it? Because the Christ-event is accessible to us only with the aid of this discursive material that derives from it, and that we must take account of, for better or for worse. In its reality rather than in its possibility, this event is unintelligible for theology; but we have new scientific means to render intelligible, as far as is possible, the reasons for this unintelligibility, and to safeguard it as a secret. It is the Christ-event that must found—in a scientific mode, not that of belief— the science of Christianity as formation of thought and of knowledge, it being understood that this scientific transformation of Christianity is not particularly a transformation of signification or of semantic exegesis, nor especially one of the deconstruction of its texts. This phenomenon is none other than that of the emergence of messianic faith and of its payload of rigor against Greco-Judaic and Pagan beliefs. Christ did not have to give rise to a religion, but the religions that encircled his cradle to welcome him like "bad" fairies did not hesitate to pass the virus of religion on to him, and to give him a seat in the war room of monotheisms.

How then, it will be asked, can faith "in" this event count as a scientific cause or foundation? But has faith ever been the belief in an event, however mysterious or objective? Is it not rather an immanent praxis, the messianic practice of the world that finds some affinity with an affect of immanence which is that of scientific knowledge as lived, not as object? We still do not know, after two thousand years of theology, what faith is, even if we all "have" it, even if we all "are" it without having a true idea of it or any cognizance of it. We do not have faith in faith, nor faith "in" Christ but, really, faith "in" (according to) Christ, or "informed" by him. Faith is a scientific-*type* principle (neither positive nor transcendental), which renders Christianity intelligible while undoing it as belief; it is a knowledge of whose unintelligibility we no longer have any cognizance. There is unintelligible knowledge, it is the cause in-the-last-instance of its own knowledge as unintelligible, and because of this it is the means of Christianity and its sources. It is not at all a matter of a new interpretation of the Christianity that is given and that we all know well, but of a Christ-science of Christianity.

Let us proceed to a new distribution in a gnostic spirit of the governing terms at play in these sorts of problems. On the one hand we radically (that is, unilaterally) separate faith, as scientific but lived principle, from belief, which inhabits religious and philosophical transcendence. On the other hand, if faith is on the order of a knowledge, it is a non-Platonic knowledge, which cannot be defined as mathematical or transcendental, but only as generic and immanental, proceeding via quantum means, with belief being rejected onto the side of illusory cognizance or of representation that is founded in-the-world. Finally, this knowledge is not self-cognizance, it does not know itself reflexively to be itself, and all the efforts of this science consist in placing religious discourses at the level of their incompetence, or in making them concretely recognize their irrelevance to a cognizance of faith. It is not a matter of recognizing that we do not know anything, just that we do not know anything through philosophy, representation, or historical belief as to the knowledge that we are generically qua faithful in-the-last-instance. It is not Socrates who awaits us at the end of the search, but the gnosis that speaks the "negative" truth of Christianity, with its occasional aid.

The counterproof of the axioms used before for idempotence and superposition is that an organizing transcendence like that of God and his plan converts simple generic duality into a triad, by making it lose its consistency, or converts our discourse into a "new" theology, destroying its superposed axioms. It is enough to introduce transcendence as a principle for humans or the simple to find themselves embarrassed, caught in those nets that are called Jewish paradoxes or Greek aporias, along with their specialists—priests on one side, philosophers on the other, psychoanalysts in between, all the great commentators and exegetes. Every historico-analytical approach to the logic of messianity purely and simply destroys the real, if not the reality, of Christ, by making his lived being of messianity or of superposition vanish.

Still, the refusal of philosophical or theological representation is not the refusal of all discourse. The Christian science is spoken through axioms (we shall come back to this) that utilize concepts by requalifying them as primary terms, precisely through the suspension or neutralization of their philosophical sense, or even as prior-to-first terms or oraxioms, thus resuming the christic operation on the plane of the word. Non-Christian science is a set of "oraxiomatic" yet immanent statements, deployed infinitely like the phases of a flux. They are neither cataphatic nor apophatic, and they cast off all theology, whether positive or negative. There is obviously nothing scandalous in the faithful assuming the operation of Christ, in speaking of axioms that are generic phrases of messianity, "messianic wave functions," oraxioms in which the generic subject expresses itself.

The Cross and the Resurrection are a secret open to humans rather than a mystery in the hands of the Church. It is important for the non-Christian science, as it is for every science, to demonstrate the limits of the validity of its proofs (of oraxiomatization), unlike foundationalism and religious fundamentalism. And consequently to be able to speak, not to be obliged to shut up but just to know how to "weaken" one's discourse—this is the function of the oraxioms in which it expresses itself. The Cross is a sealed secret, fulfilled in the person of Christ, accessible to the Simple Ones who *are* unlearned knowledge, but refused to the sufficiency of theologians and philosophers, who know that they know nothing rather than not knowing what they know. Their presuppositions automatically imply the destruction of science, for one cannot decently call "science," except in appearance, something that is founded on philosophical means and procedures.

Christ is a "real" thought-event: he determines, and transforms into his object, received thought; he "fulfills" religious discourse. The fulfillment he claims cannot have any status other than that of immanence or of the Same, and can only be fully understood through idempotence. He no longer engages in the relations between man and Being or Logos, still less those of man and the Torah, but only in those of man and the world insofar as humans have to generically work within it.

THE SYMPTOMS OF GENERIC SCIENCE IN CHRIST

It cannot be a matter of a classical, positive science, a science almost without out a subject—nothing here corresponds to such a thing. Jesus can perhaps be said to have conducted a laboratory experiment, but not in this positive sense; it is a kerygmatic experiment, carried out in the cramped territory of Judaism, an experiment of salvation in the vessel of the world. All the same, in Christ's practice and above all in his sayings, we find many ingredients for a science, but it is a science that includes the subject, in the form of symptoms to be analyzed, on condition of knowing how to read the principles that we "have" a priori in these symptoms. We know the symptoms: the sense of universality as human rather than as cosmic, its fundamental relation to the Law that remains the Law even when fulfilled, the immanence of its fulfillment "in" his person, the theory of his new mediating function, his doubly irreflexive thought—on the one hand his simple, inaugurating, and definitive sayings, like axioms, which, on the other hand, are addressed to "simple souls," at the limit of the "simplicity of spirit," who know nothing of philosophy, the necessary passage through the literally crucial experience of sacrifice, and finally the affirmation, accepted by the Apostles but suspect in the eyes of the Church, of the actuality of his messianity and of his accomplishments. All that Christ says under the name of *fulfillment* is obviously capable of giving rise to religious scenarios or dialectical constructions, but can also indicate in outline the enterprise of a new comprehension and being, to be taken literally as actual, as axioms in which Christ strips himself of all attributed religious meaning, without necessarily operating a "negative theology." It is up to us to receive them in a nonreligious and ultimately nonfundamentalist manner.

All of this is not merely existential, and is not, moreover, actual in the sense of historical existence deliverable into the hands of theologians, if not sectarians; but it is theoretically relevant if one knows how to decipher it and see its logic. Messiahs are only actual under the virtual condition of messianity, which includes a trait of inexistence or of nonmanifestation, signifying that they do not manifest themselves to the eyes of the world, which is blind by definition. Theology, cosmological in form and in vocation, philosophical in its means, has constructed upon the event of Christ-thought a whole "plan of salvation" attributed to God, a vast story that makes up the ground of our mythology and that is pursued by teleology in the direction of the history that is supposed to encompass humans. But a non-Christian science, founded on algebra, the algebra of idempotence, and no longer on a philosophical logic, will instead grasp in messianity an almost aleatory process, an indiscernible dispersion of messiahs sufficiently intricate to outline the "ecclesial" body, but nonlocalizable by the faithful. To pass from theology to the generic science of Christ one must pay this price, which is not the abandonment of all "hope"—quite the contrary, perhaps.

THE QUANTUM MODEL OF CHRIST

We transform Luther's still theologically inspired imperative by adjoining to it the concept that it lacks; we reinstate its insurrectionary status as "philosophizing in-the-last-instance according to the science-in-Christ." The West is sufficiently proud of Christianity's invention of another conception of man, of another relation to the world, of the introduction of the individual and the subject into history, and even of the birth of modern science. It is time to draw all the consequences of these affirmations and to give them some scientific rigor. If Christ inaugurates a science of humans qua submitted to the world rather than to Being, then we must go as far as possible in the most innovative current science of the relations between man and the world, and, for this new impetus, turn to quantum physics. Hegel, in his speculative way, had probably recognized the characteristics of a possible science in the sayings of Christ, but he practiced it in apparently philosophical ways. We take up this problem again in a non-Hegelian manner, neither positive nor transcendental or speculative. The possibility of this previously impossible treatment owes to the existence of this already secular physics, which furnishes an entirely new way of reading the experience of the world, and thus a new way of thinking to rival philosophy. Quantum physics will serve as our model to transform the style of thinking and even the practice of faith. This is why Christ will be treated as a "christic constant," faith as a "quantum of faith," and the theoretical procedures implemented as mathematically stripped-down quantum-theoretical principles, the two principal among them (obviously interrelated) being the superposition characteristic of wave phenomena and the noncommutativity detectable in particulate phenomena.

All we are doing is taking up again, in our own way and a little better armed, Christianity's ambition to conjugate faith and science, but in a way that owes nothing to the confrontations or syntheses operated by self-declared or nostalgic atheists. This science is not historical Christianity itself, and in general it is not a religion; it is the science of existence, whether narrowly "Christian" or otherwise, in the world as Christ unveiled it or posited its foundations. If Plato is the emblem of philosophy in its twin-ness with mathematics, then to philosophize-in-Christ implies a displacement of philosophy and theology placed under determinant condition that is no longer itself Christian, but christic. Under condition, as would be a Last Instance that would seem to retire or subtract itself from theology and from religions, producing in them an effect of the underdetermination of belief. All the same, it is not a simple subtraction operated *on the sly* in the name of a faith that is itself vanishing, and would believe that in this way it could smuggle itself past any theological control. Radical faith is prima facie conceivable as a quantum-type act of superposition, and is opposed point-by-point to the identification of belief that would always seek the absolute. The quantum of faith is a sort of nonacting or nonreaction to the world, but one capable of acting by transforming the world without, therefore, creating it. Correlatively to the irruption of faith, we must abandon the model of creation that is not only "objectivating," but philosophical or theological, and what remains of it in the opposite, specular model of its idealist or materialist contemplation, in favor of the generic model of its transformation. Radical faith, the fidelity-in-the-last-instance permitted to humans, has an immanent effect that alone is the generic transformation of the world.

The fusion, in the name of "Christ," of scientific principles with ancient religious beliefs both transformed and put into a quantum state of superposition prohibits any kind of return of philosophy upon the messianity that is foreclosed to it—that is to say, a twofold traditional enterprise. Firstly prohibited is a religious, ultimately circular and philosophical critique of philosophy. To philosophize in Christ, this slogan has been perceived as a way to bring down the grandeur of Reason by way of Plato—the whole Lutheran tradition that runs through Kant, Fichte, Jacobi, Hamann, and Kierkegaard. But this is not at all our project, which is not intrareligious and antirational. Even if we are happy to be neighbors to these thinkers who struggled against the great rationalism of dominant philosophy, we interpret the formula on the basis of principles both philosophical and scientific, and this in quantum thought and in Marxism. These two disciplines serve us as models of a science established upon these principles, models in the axiomatic sense. Subsequently, and symmetrically, the theme of fusion with a subject become quasi-predicate shows that it is not a question of a scientific-positive *foundation* that we wish to substitute for Christ, and that would flatten everything onto a positivism without being able to give this science a real dimension that could be said to be "immanental." Strictly speaking, it is a matter of superposing, not identifying, the grandeur of Reason and the religious philosophy of the Greeks and Jews who tried to capture and produce "rationalism," or conflated this grandeur with philosophical sufficiency and scientific positivity. In all seriousness, there is no grandeur for this christic and generic science as there is for philosophical Reason, no Christian theology as rigorous science nor, inversely, science as rigorous philosophy—only the indivisible bloc of a *christic subject-science* whose object is Christianity and its theology. What must be thought, or rather what makes thinking happen and constrains reason as much as it constrains belief, is the undivided encounter of certain scientific principles with a subject plunged from the very start into a scientifico-religious doxa from which it is extracted, constrained by their radical immanence to become a faithful subject. A faith is necessary to practice this science, but does not condition it; on the one hand it is a negative or underdetermining condition, on the other hand faith is belief transformed by science and desubjectivized, neutralized, or made generic. Our beliefs are philosophically bipolar, individual, and thus gregarious, whereas our faith is generic. We do not have to bear this faith either as individual or as collective. To philosophize *in* Christ but to believe *in* one's capacity as scientist—this formula may be too paradoxical for positive spirits to understand, except under a certain number of quantum conditions including that of superposition.

THE EXPERIMENTAL SCIENCE OF THE CROSS

The Cross and the Resurrection, which are inseparable, are the heart of Christianity, its genetic kernel as religious experience and as faith. But they have been interpreted rather hastily, from within a bastard Greco-Jewish mélange, rendered unintelligible beneath heaps of images and beliefs, transformed into rationally incomprehensible paradoxes that call for the easy solution of substituting belief everywhere for faith. Belief and representation have served to efface the blank spaces of theology or, inversely, theology has plugged the holes of belief. For us it is not a matter of trying to "rationalize" them in the classical manner, as idealism claimed to "rationalize" experience (Kant, for example, with his "religion within the limits of reason alone"). Rather than these mixtures, how can we invent a generic practice of Christ and of messianity with quantum means, means that therefore are still scientific, but better suited to the *contemporary* character of Christ? For he is our contemporary, and we are the contemporaries of his messianity insofar as his immanent act ceaselessly demands new theoretical means, and must throw off the old transcendent finalities (classical and modern alike) with which history has loaded down its messianity so as to divert it from its effectiveness—namely, the "salvation" of which humans are capable. It is with this intention that we construct *a scientific model (not a "representation") of the Cross in the form of a generic matrix* that is more manipulable, and with more certainty, than theological images and beliefs. It is capable of bringing to light the generic truth of Christ in the very act of "crucifying" philosophy or the world. For historical Christianity and its theologies are, precisely, a representation of the messianity of Christ, just as belief is a representation of faith.

The theo-christo-logical doublet is organized by theo-centrism, and on the theological plane by ecclesio-centrism. These are the fundamental characteristics that we want to destroy here, but without carrying out a textual deconstruction in the Judaic spirit or a neotheology that is still in the spirit of the Fathers. We force theology to change its theoretical basis: it will no longer be based upon (essentially Greek) philosophy but upon a science, quantum physics. By substituting the principles of contemporary physics for the old classical rationality that was in reality entirely philosophical, we do not eliminate all recourse to philosophy despite its being relegated to a secondary or, as we shall say, "occasional" function. In order to achieve this we shall do as the theologians do, concentrating upon the meaning of Christ at the moment of the Cross, but so as to draw from it entirely other consequences. It is a matter of arriving at an interpretation of the set of phases of the sequence *sending of Christ/message/death/resurrection/evangelization* as the unit of a crucial experiment of humanity, with preparation of the experiment, sacrifice, proven result, and objective appearance of the phenomenon. The texts of mystics could also have served us here as a guide for describing the *experimental science of the Cross* as lived experience. But we gather them together and treat them as and *in an experimental vessel* entirely distinct from the temporal dispersion of the historical story—of evangelization, for example, as the history of the expansion of a belief.

FAITH OR FIDELITY AS GENERIC QUANTUM

To delimit the human phenomena accessible to the new science that Christ discovers, it is fundamental to identify, alongside the experimental vessel of Golgotha, a constant, a generic quantum that assures us that we will be dealing with generic humanity and not, let us say, "rational" or "logocentric" humanity. This will be faith or fidelity qua identical to messianity, despite (as we shall see) their duality with the latter. There is no rigorous discipline of religions that does not seek to be positive, but usually they do so with physico-mathematical constants that swamp humans with anonymous and overgeneral entities. Their place is then taken by philosophies and theologies that are not sciences. It may be that religions have affirmed their sufficiency, and that of their theologies, upon the basis of this absence or this theoretical repression of a generic science of faith. The disciplines that exist use positive means (which can be philosophically elevated) such as history, archaeology, exegesis, even dogmatics, but without a generic means, without the subject that is implicated in them. So that the faithful subject is presumed sooner or later to be a philosopher, and therefore a believer. These objective sciences of religions forget that the true constants, as universal and rigorous as they may be, are also specific to their object, here "the simple ones," those stripped of the Principle of Sufficient Philosophy, but not of all philosophy. This is why these two determinations of every generic human, or of what he can do, are conjugated in science qua generic. We have also seen a deconstruction of Christianity, psychoanalysis, and structuralist enterprises, but from our point of view all of these rest either upon scientific and "textual" ideologies such as structuralism or (like psychoanalysis) upon a basis of positive sciences (biology or mathematics) absorbed into a Judaic context. For this reason they could not do justice to the subject qua generic—that is to say, qua inextricably faithful and believing—in such a way that one would be able to say that humanity is faithful in prior-to-priority, believing in priority.

If we conceive the human constant of the science of religious phenomena as faith or fidelity, it is a matter neither of just any "humanity" nor of just any belief, but of the generic concept of the subject = X to be determined as the support of the conjugation of faith and belief. A constant necessary for all human, political, aesthetic, or theological sciences, which could, for example, make of "theology" a human science of God rather than a divine science of man. This constant must be discovered as the scientific threshold to be passed through, a threshold that opens onto a region of unlimited possibilities. Prior to the discovery of this constant, a science works above all in the imaginary of its object, does not distinguish itself clearly from this object, and loses itself in mélanges. This constant can no longer be of an institutional nature, as Catholics would have it, or scriptural, as would Jews and Protestant Reformers. We no longer have to choose between the dogmas that structure the Church as thought, the Scriptures whose consistency is reduced more and more to a fundamentalism, and a sectarianism just as closed as the Church—between two deadly dogmatisms.

Obviously only the sciences can introduce a constant that would not be a philosophical or theological subtotality. But the constant, as objective as it might be, must always also imply a quality that belongs to the domain of objects. Even the quantum of action or of the speed of light contains symptoms of a subject that will reveal itself subsequently—for example, in quantum physics, in the form of the "subjectivity" of measurement or observation. As for religions, what is specific to their object is the believing subject. One thing that is original in Christianity is its having revealed the generic essence of sciences that claim to be human and that must engage immanently, in their scientific stance, with a lived that is still more faithful than believing. We draw the consequences of this in the transfer of the science of religions into the quantum terrain.

NAMES AND FUNCTIONS OF THE RADICAL "CHRIST"

The science of Christ is not that of Jesus. The name "Christ" is a volatile hotbed of theoretical functions—unlike Jesus, who is more easily discernable by the historical and theological sciences. It is impossible for us to speak of "Christ," period, in a unique sense, except obviously (and this will often be the case) in simplifications, generalities, or quick references. We distinguish between a diversity of functions into which the "radical" Christ enters.

Christ-System

Open set of utterances of every kind in the Gospels, whether canonical or not, the logia of Jesus, the tales of the disciples, the interpretations of the Apostles, all of which relate themselves exoterically, in Christianity, to the pole of Jesus Christ, but which function here "esoterically" as properties or variables in the science of Christ, in view of determining the knowledge that we can obtain of the messianity of Christ through faith— that is to say, of the knowledge of Christ-in-person. The kerygmatic content of these utterances or "sayings" is thus suspended in its theological sense along with the Principle of Sufficient Theology, but not destroyed: reduced, rather, to the status of properties of Christ that can be treated as variables in a device called the christic matrix.

States of the Christ-System and Messiah-Function

Designating the different possible combinations of these variables to which the system gives rise; with each state of the system containing the traditional or theological Christian variables of Christ, or his Jewish or Greek coordinates *plus* the messiah-function, which is added to the variables and permits a quantum-theoretical treatment through its algebraic character as imaginary or complex number. The states of the system are formed of various (theological and scientific) cognizances, but are not yet faithful cognizances; they are the variables of Christ as real object of cognizance, but only faith as faithful cognizance of Christ is the object of cognizance properly so called. We prohibit ourselves from confusing the data that make up the real object with the matrixial process of the production of faithful knowledge.

Christ-in-Person

We distinguish the messiah-factor from Christ-in-person as such (*tel quel*), qua incarnating the messiah function, and Christ-in-person itself (*comme tel*), qua object of cognizance. This is the unity of messianity and the clone as the two wave and particulate aspects in their unilateral complementarity. Whence the ambiguity of the In-person, which is the indivisible Christ and his unifaciality or his Stranger-subject-being. It is Christ complete or as duality grasped according to messianity but also existing in the state of a clone and consequently open to the ambiguity of the clone-Christ and of Christ in-itself, on whose basis one may go in both directions. Whence the difficulty of defining who Christ is, his probable ontology: outside Christianity or within it? The In-person is characteristic of the final result of the matrix, or the faithful cognizance of messianity that it produces.

Messianity and Faith

Very general common designations or notions that relate to Christ and hence to the faithful-subject insofar as both of them are real objects and objects of cognizance, Christ and cognizance of Christ. Hence the extended range of uses we make of them. The essential distinction is that messianity is the generic lived phenomenon (qua real or qua knowledge), and faith or fidelity the form of the clone of the faithful subject always on the edges of belief.

THE MESSIAH-FUNCTION 1: INTRODUCING CHRIST AS IMAGINARY QUANTITY OR INDEX OF SCIENTIFIC FICTION

To locate a contemporary scientific level in the domain of theology, we must accept the introduction among its givens of a factor characteristic of physical and quantum scientificity. Alongside the real but historical Jesus Christ, Christ the religious object of theology, there is another Christ that we could call "imaginary" or, paradoxically, "scientific" because of its proximity to the algebraic or complex imaginary number—a messiah-factor, more precisely, charged with making the cognizance of Christ pass from the state of a body of theological knowledge to its generic (that is to say, scientific and more particularly quantum) state—truth. This is the famous "spark" of the gnostics and mystics, the "pearl," as a last radiance, the last-instance radiance of the flash of the Logos. There is a function of the messiah that must be compared to a purely algebraic datum of Christ. This radical Christ should be imagined on the algebraic model, imported from outside so as to secure a possible science, a rigor of thought, not drawn as a given from the Christ-system of the Gospels, but preferably from the power of human understanding as original factor that shakes up the sufficiency of relations or of the theo-christo-logical doublet. A real Christ in a new, nonrealist, thoroughly messianic sense will come forth in it.

On the one hand, like all theologians themselves, and above all in view of the principles of a science of monotheistic religions, we are obliged to suppose that in general they are of a rational nature, at least in-the-last-instance, that they contain a fundamental spark of rationality accessible to a science, which we aim to find. Science in the original Western sense—as distinct from those mythological thoughts or religious wisdoms that escape our power and our knowledge, however accessible they may be to other rigorous forms of knowledge apparently more suitable for these practices (the human sciences, structuralisms, psychoanalyses, cognitivisms)—a sense that does not exclude our attempt to include religions in physical nature.

This factor that we shall call the "messiah-function" or the "index of christic fiction" is apparently present in the theo-christo-logical doublet, and must be drawn out of this structure. No doubt it must provide an explanation for the latter, but it must first descend from the theo-christic doublet toward the messiah-function, subsequently to return from the latter back up toward the doublet. This return is the role of the matrix that produces faith in Christ, the final object of our experiment; but a prior movement is necessary, which is not yet a matter of science, but of its preparation and the inventory of its conditions. Given the doublet, it seems impossible to obtain the messiah-function in its exact figure because one cannot then exit from the doublet, and from the double transcendence that coincides with the PST. This is the vicious argument that Kant thought he had been able to break through. But despite its modifications, extensions, flexibilizing, a science supposes in its object that which it itself has essentially put there a priori—a form of rationality that is adequate to it. A quantum-oriented science places in its object at least the great rational principles that belong to it, places them in it in their totality, without being necessarily positive for all that. For *the quantum a priori* comprises the three principles of superposition, noncommutability, and entanglement or nonlocality, but also what we shall call, provisionally and "in Kantian," the *transcendental a priori* of these principles—namely, that spark that is the complex or imaginary number ($\sqrt{-1}$) that is necessary to enter into the sphere of quantum pertinence and to render these principles applicable to religious experience.

As a counterproof, it is possible to deduce from the theological One or from monotheism (that is to say, from a theo-christo-logical doublet structure) the messiah-function as its radical origin, or even the imaginary number as kernel of Judaism. We must first distinguish cognizance as datum in science or outside philosophy, and that same cognizance as taken up in philosophy—this is the level of Derrida's deconstruction, which presupposes the brute or linguistic signifier and then goes on to grasp it again, philosophically.

We posit the messiah as the non-"real" factor, in the sense that it is nonarithmetical but algebraic, as what is algebraic in relation to theology or to religion, an unknown recognized as such by the Jews (and the Christians). Rosenzweig appeals to a scientific "parable," a comparison with mathematics, so as to demonstrate the irreducibility of the Jewish people qua exception that does not "number" among the nations, a "nonnumber." The irrational number concretely manifests the infinite, its alterity and its transcendence, become visible and physically present while remaining foreign in reality, incommensurable but physically present, revealing an absolute infinity that surpasses the measure of sense, whereas for the rational number, the infinite is a certain but unattainable and abstract limit. This opposition comes down to establishing a Principle of Sufficient Judaism (PSJ) against the Principle of Sufficient Philosophy (PSP)—two different types of sufficiency and an opposition of the presence of the Absolute and the infinite of the imagination. We must deepen not this opposition of the Greek and the Judaic, but the more complex opposition of the latter with the imaginary number. The imaginary number is also no longer a "number" in the arithmetical sense, and is measured less against the absolute than against the radical that it bears with it and introduces into thought as a prior-to-priority stripped of foundation and contingent in a henceforth nonphilosophical sense.

In other words, our parable or our fiction is precisely that the name of Christ both designates the equivalent of the imaginary number operating as the "puzzling" Stranger or messiah-function in the Greco-Jewish or Christian milieu and introduces the power of fiction into it. The variables of the Greek and the Jew are multiplied and affected by the idempotent reprise of the Jew or of the Messiah, who apparently returns, but who in reality comes for the very-first time, who undercomes or sub-venes (*sous-vient*).

The Christ-factor exposes or "has faith" in the Resurrection rather than in the Crucifixion. The Resurrection is a meta-physical act—that is to say, a meta-quantum-theoretical or generic act that establishes a messianity more complex than Judaic messianity or than a Greek-style intellectually biased act. It gives us the measure of an indiscernible Christ, and in doing so distinguishes him from Jesus, who falters on the Cross. Can we not say that the death of Jesus on the Cross is itself resurrection-oriented, oriented toward Christ as generic figure of Jesus? It is vital to suspend the Logos and its transcendent anthropology and its mathematico-Platonic philosophy, so as to bring back into play the spark of rationality that it contains as physics. We interpret the notion of a messiah-factor upon which to index the christic science as mathematical and algebraic, but relieved of the PSM—this is the minimum requirement for the quantum orientation of the knowledge of Christ.

The Christ-system thus comprises two closely conjugated aspects, and only two. On the most external level it designates the christic message, the basically Greco-Judaic kerygma, along with the set of utterances, formulae, and logia of a character named "Jesus," and, with this, the whole of Christian theology that envelops him as theo-christo-logy. But more profoundly "Christ" is the symbol that designates a structuring but immanent function of those messages, an essentially quantum-type fundamental variable, a factor that is added to all of the variables extracted from the messages, and that has the power to transform them into vectors. It is a factor of the vectoriellization of faith, torn away from belief and even from the imaginary fictionality that gives the sense of its message. Christ plays the role of a symbol for a quasi-mathematical or algebraic operation; as that factor, he expresses and determines the essence of non-Christianity or nonstandard Christianity. Unlike the Will to Power factor that is added to forces in Nietzsche-thought, that is to say, in the essence of philosophy, to further potentialize it, Christ is the factor that weakens or underpotentializes all messages that are conducted by the PST.

Thus, the content of the matrix is not solely Christian in the narrow historical sense; the Christian is an after-the-fact construction of messianity or of faith on the basis of Greek philosophy and the messiah-function. The latter is not entirely the same as Judaism or the Law, the Torah. It is added to the Torah just as it is added to the Logos. It is more universal than in its Judaic understanding, and refers to the theo-messianic or theo-christic doublet that corresponds to Christianity.

THE MESSIAH-FUNCTION 2: THE INTUITION OF THE CROSS

These preparations having been made, a problem remains in suspense. All of these principles are scientific, and specifically algebraic-style, cognizances. Have we fallen back into a philosophy of science, a quantum ontology—and, worse, a transcendental one? Not at all, or at least not entirely. This interpretation, which one might call transcendental, is only a phase of the preparation or of the internal workings of the matrix. But if the latter must be quantum-oriented, we must find the conceptual equivalent of the imaginary number that is closest to its mathematicity and its quantum usage—the equivalent that, as conceptual, demands that its schematization in mathematical intuition be the shortest and most inevitable for these two idioms—the concept and the imaginary number—to be able to communicate or coexist adequately. They have a common element that will be represented by the circle and its division into quarter-turns, as we see in the Cross wherein it is visible. The Cross is more than a historico-spiritual symbol, it is *the christic intuition, of a quantum nature*, in which the theologeme of Christ and his human figure communicate.

This intuition of the Cross is more than a transcendental a priori for the science of Christ: it is first of all a mathematical datum, a brute and primary algebraic given, but one that is not reprised or interiorized in a simple Kantian transcendental gesture, or perhaps in a geometrical schema (itself an object of mimesis, at once the gesture of a believer and a painting that the painter shows to us). All the same, we have not yet left the transcendental sphere, we are wandering on its borders. Another step must be made beyond the image and the symbol of Christ on the cross, beyond this schematization that thinks the (hypertranscendental) messiah-factor negatively, abasing it by way of the pictorial or imaged means of the concept of Christ, a concept still attached to the cross of philosophy from which it hangs. What is missing here?

In order to make our way to a positive reason of the Cross, one capable of exposing its genesis or explaining it, what we are lacking is the entirely other interiorization of the transcendental itself and of its double transcendence within a more real element—that is, the imaginary or complex root of the schema, which left to itself repeats and redoubles, potentializes itself philosophically and thus enters into contradiction with signification. *The Cross has not itself been thought crucially by theology, it has not been reduced and engendered by the quarter-turn that is the root of the circle.* What is lacking is the leap from a philosophy-of-quantum-physics to a quantum-physics-of-philosophy—that is to say, the possibility of the cognizance we seek of Christ and of the Cross, but of which, for the moment, we have only the preliminary conditions, since we still conflate the real object with the sought-after object of cognizance. We remain in the exteriority of a transcendental operation of schematization—that is to say, in the transcendental imagination. No doubt we could announce, oversimplifying somewhat, that if the algebraic version of Christ is the imaginary or complex number, then Christ is the religious version of algebra. Christ would be the algebraic reduction of religions, the formal supplement added to every possible religion in order to orient it generically. It is this that would endow Christianity (however overrun it is by pagan, substantial, and even monotheist dross) with the character of a formal, rather than substantial, religion. The reduction of monotheisms is demanded by the christic factor, the ultimate reduction of all substantial and theological content.

Still, the problem was how to reconcile the brutality of algebra, this datum of the complex number, and the meaning, the flesh, that the schema implies. It is here that quantum physics is no longer negotiable but imposes itself all at once: superposition alone responds to their superposition. There is (and this is more than a datum, more and less than a factum) a superposed-without-superposition of the concept of Christ and of the geometrical (or in any case intuitive) flesh that still served as its schema; with their fusion, the schematism is negated in its hierarchical and philosophical form. The Cross is not, and obviously cannot be, self-caused, on pain of an ultimate nihilism. Certainly one might imagine that death is proof of death, but the lived explains it still better—that is to say, by quantum procedures. Its positive cause lies in the superposed state of Christ and the resurrected Christ. Obviously the secret of the Cross can only be the resurrected Christ, since it alone manifests their intuitive common-being that is elevation or ascension, the vectoriell ascending that is its reprise. It is the Resurrection that explains the cognizance of the Cross, not the other way around; the death of Christ proves almost nothing, it is his resurrection as insurrection that is the true ascending that, as a repercussion, abases the theological concept of Christ, and theology itself. The death of Christ on the Cross is the death of theology, which has been crucified and which is resurrected in faith. The messiah-function is Christ resurrected, that which transforms Jesus and his sayings into an object of cognizance, but which cannot be *deduced* from them on pain of returning to a mortifying theology, and which is commutable neither with them nor, more generally, with Christianity. Not to mention that what is to be discovered here is also the inversion of Schrödinger's paradox.

CHRIST AND THE ALGEBRA OF RELIGIONS

Christ was swiftly rejected onto the margins of theology as a simple means or mediator, or even as a religious persona—sometimes as a false messiah, or a fabled character like Buddha or Socrates, a master of wisdom. Or a Jewish sectarian. Or a historically dubious character, and affirmed all the more dogmatically for it. Faith requires a more solid (even if strictly speaking indeterminate) reference. An algebraic interpretation of the Christ-function is necessary. Unlike Jesus, who gives rise to a transcendent and theological imaginary, uncontrollable and therefore demanding to be dogmatically standardized and forced, under these new conditions Christ gives rise to a controlled and limited, generically oriented fiction. The nonecstatic orientation of vectors is what remains of simple (not reaffirmed or redoubled) transcendence; immanence, albeit radical, like a simple ascending, no longer takes place in the absolute (the dis-oriented milieu par excellence).

The Christ-factor must be understood as a negative and irrational concept, like the flight of the Angel: an ascending or unifacial vector, a principial vectoriellity. Christ's sayings are not behaviors or acts, but vectors or throwings, ejaculations. The Russian mystics knew how to decipher the meaning of these ex-pressions. Certainly faith is not a cry, but neither is it a dogmatism. Christ is himself a resultant of vectors that traverse theological transcendence. Faith is an ascending or a vector with a minimum of the particulate, or a kernel of belief that is the phase, whereas faith corresponds to the flux. So, no faith as pure interiority of consciousness, but a unilateral complementarity of faith as modulus and amplitude and belief as phase or particulate direction. This is another way to think faith as an ascending stripped of all macroscopic ecstasy but not of any "object." Apophatic and cataphatic are to be interpreted vectorielly: the vector is at once cata-phatic (because it departs from a radical and generic origin and declines), and apo-phatic (because it ascends toward a particle of faith).

QUANTUM THEORY OF THE BODY OF CHRIST

The human incarnation of the name of Christ must be chased out from its last idealist and theological retrenchments, and completed in the form of a physics of the christic body. What would incarnation be, relieved of its mythology and its beliefs, delivered to the faith of the faithful, if not a quantum-oriented physics which permits the deduction that we must call (despite the apparent absurdity of such a formula) a microchristics or a microscopic christology—that is to say, *one whose intelligibility can only be vectoriell rather than existentiell?* The archaic character of theology, positioned on the frontiers of religious mythology, can only aggravate philosophy's position at the frontiers of science; but this is no reason to be disinterested in it, in the name of an atheism that has little to recommend it.

Let us remain for a while on the borders of meta-physics—that is, for we the contemporaries of "our" sciences, the borders of meta-quantum-physics. We seek a physics of the body of Christ, and abandon the psychology of belief. That Christianity introduced a dominant notion of the "person," that it is a matter of this Christ here who was crucified for us, one time for all, and who came forth as Christ-in-majesty, these are all just images or consequences, and not yet the generic concept. Let us use philosophemes at least as a provisional aid, subject to a complex or more rigorous (more precisely, inseparably theo-quantum-theoretical) understanding of them. With Christ it is a matter of what we have called the

"One-in-person," certainly not of Being, or even of God, or of that One of which the mystics were the consummation but which still lingers on the desirable margins of the Logos, both distant and too close. There is no longer any place here for the weighty dramaturgy of the Trinity, except insofar as it presents itself to us as a symptom. It is enough to understand that "in-person" is the result of the "superposition" of the One and of the person, of the "in" as in-One or immanence, and of the face of the One or the mask of this immanence. Now, this superposition—which is no longer a philosophical operation like the Parmenidean "Same," but physical, or at least quantum-oriented—can only be said of two vectors, and certainly no longer of objects or concepts and metaphysical categories. With this term "superposition" we have already transformed all of christology into christo-fiction. "In-person" is the unique face of the One according to a relation-without-relation of unilateral complementarity. The face of the One does not add anything to the One except itself, but *the One adds immanently to itself the face or the person that is added to it*. Obviously, under the macroscope of philosophy (which is "decoherent," as certain physicists say), the formula loses its superposition, and is decomposed into an amphiboly or a specular doubling; it bilateralizes or becomes reversible, as in the famous axiom of the convertibility of the One and of Being.

Three

FROM THE THEO-CHRISTO-LOGICAL DOUBLET TO UNILATERAL COMPLEMENTARITY

THE THEO-CHRISTO-LOGICAL DOUBLET AND THE INVENTION OF CHRIST

Christ is said in many senses, and remains an equivocal name. Rather than adding yet another sense, we seek to give it the function of symbol for a scientific event that opens up a new field of research, and is capable of treating problems of theology—instead of doing the reverse, explaining for example why it bears these multiple significations. Theology has always claimed to be a science, but in a philosophical, Aristotelian sense, not in the experimental sense, which is at once more rigid and more supple, according to the style that modernity has established for it. Now, a science begins by cutting out its object so as to acquire a more rigorous rapport with it than that of common belief and theology, or even philosophy. "Christic nontheology" proposes to cut with the weapon of faith the theo-christic doublet that structures Christianity and engenders its ecclesio-centrism. The correlation of Christ and God is the object of an imaginary and religious science that has appropriated it and institutionalized it, a Greco-Judaic confusion that must be undone. Christ has given rise to an interminable exegesis, but we make Christ the name or the symbol of a science-in-person of religious phenomena. It is important to break all imaginary or theological continuity in christic causality, but without pronouncing their absolute incommunicability—that would be the philosophical temptation, and it would be a specular illusion. A science of theology breaks its bonds with theology without nostalgia (a decisive point), but brings with it, in its baggage, a certain part of theology that it will make use of as one of its means in the service of the scientific cause, but that will not be able to reciprocally determine it. This complex rapport, in which Christ breaks his bonds with theology yet conserves certain secondary links with it (as a material of symptoms and of productive forces), defies the logic of the theological understanding. We call it the unilateral complementarity of Christ and the discourses that are maintained about him. It is unthinkable to break radically (and not absolutely, like atheism) every link with theology and philosophy so as to better leave the latter to their autonomy and their sufficiency (Principle of Theological Sufficiency). It is a matter of abasing this double and no doubt unique religious and philosophical sufficiency, and of breaking with it only so as to make a better use of it. Christ is incommensurable (but incommensurable otherwise than is God) with those theological affirmations that vainly assure themselves of him. Christology is an apparent science, a sector of theology; but there is a science-in-Christ that is the science of christology itself. It is fundamental as a priority (we might even say a prior-to-priority) to desuture Christ and the theology that has specularly taken hold of him, to make of the former the cause or the true subject of science, a subject-science ultimately, and to dismiss theology, reducing it to its function as occasional material or variable. It is obvious that this nonrelation is difficult to think, incomprehensible to the classical, and in general philosophical, understanding. But it is incomprehensible because it is "imaginary" in a new, algebraic sense, which is not entirely cataphatic. Precisely, if theology is ultimately destined to a certain cataphasis whose mechanism of depotentialization must be reevaluated, this is the task of a science whose cause, Christ, is no longer an image of God, the deficiency or inequality of a Son in relation to his Father. In the Cross and above all in the Resurrection, we find concentrated the mechanism by which a generic science frees itself from its scientific and theological positivity as well as from its christological inversion (also supposedly sufficient). In other words, the Cross is a radical but not absolute desuturing of doublets; it is the form par excellence of the unilateral complementarity that must be brought into play on both of its two sides. The sacrifice of the Cross will be interpreted as a way of sacrificing the theo-christic doublet that is specular and belongs to a religious regime, and of assuring the invention of Christ as Resurrection and Ascension. Just as the invention of Christ is no absolute or philosophical gesture, the eviction of God is no "death of God" in the traditional sense (for example, a sense in which it would signify the triumph of the "man" of humanism). The objective is clear: to destroy the theo-christic doublet in order to save the kernel of the real within it, so that it may not destroy the messianity of Christ and refer it back to history once more. And equally, if God is thought as that omnipotent One-of-the-One, then his depotentialization takes place as the immanence of the One-in-One and of its residue of idempotence that is said of all his predicates and affects them all. The specular theo-christo-logical doublet is deconstructed by placing Christ prior-to-priority, as the factor that opens up the "imaginary" dimension of christo-fiction. The phenomenal or real content of this doublet is messianity, which henceforth forms a unilateral complementarity with the faith of the faithful. It is no longer with God alone that the soul speaks, it is through Christ that it allows itself to be underdetermined. The end of monotheisms understood as the end of theo/prophetico/christo/ecclesiastico-logics.

THE USAGE OF THEO-CHRISTO-LOGY AS VARIABLE

Theological erudition—the detailed knowledge of subtle discussions and distinctions relating to the Trinity, for example—is of no fundamental interest to our project, it may just have a certain utility. We put it aside here as being without any pertinence for understanding and undoing the doublet that imprisons faith. Its secret is elsewhere, very close to the onto-theological constitution of metaphysics (Heidegger) and that part of it which has passed into theology. Metaphysics and theology, as different as they might be in inclination, object, and history, and without posing the infinite problem of which has influenced the other, have the same structural constitution in all essential respects. Philosophers and theologians will balk at this simplification, but it is important in order for us to concentrate on the sole decisive problem for us, and to place it in opposition, or at least to distinguish in it a way of thinking that would be without doublet and of a scientific essence. What does the theo-christo-logical constitution stem from, if not from a certain chance, multiple identification—an identification of the two dualities that dominate our thought, that of Being and of beings, that of God and of the creature? For reasons of experimental manipulation or "preparation" as physicists say, the traditional ways of evoking them allusively, of locating them locally, of making them move by displacing them, of raising them by sublimating them, and finally of conserving them under their theoretical modalities, all these are important to us. But it must be understood that for a christic science, such dualities, as overdetermined as they may be, have no scientific meaning, still less a generic meaning. The fact that it is a matter of dualities in general is not the problem; the problem is their specular and macroscopic doublet form. The scholars of onto-theo-logy get lost in the minutiae of quibbles and byways that philosophers practice, when they refuse to realize what ought to be blindingly obvious: that all transversality, and even unilaterality (philosophy always provides some bait), is ultimately taken up again in a second movement of transcendence, a reversed or reversible unilaterality, completed but at the same time closed or reclosed, looped in a double turn on a globally transcendent(al) model. Transcendence always has the last word in these supposedly immanent affairs—a double transcendence, which innervates the transcendental through which it relates itself to experience, and prevails over it as the creation of the world prevails over its constitution. The problem of philosophy and theology is their naivety or spontaneity—they ceaselessly divide, distinguish, only to let the divisions close up again and become encysted in a unity that returns or that always comes back beneath them. They imprison themselves in an appearance that they practice without seeing—appearance and sufficiency. In other words, unilaterality does not exclude duality, but it must fuse with a milieu of radical immanence that we have always called the One-in-One or vision-in-One, on condition of their vectoriellization, and deliver itself from the enclosures of transcendence that are so many reclosures—a double, hardly a multiple, an indefinite hardship for humans.

The theo-christo-logical doublet is, however, a correct provisional posing of the problem of Christ, even if it is as yet only a presentiment of science. The argument that the transcendental is only cut out or detached from a more encompassing "metaphysical" sphere is fruitless, for one can say nothing rigorous about this sphere without this detachment—one can only prosecute eternal wars, like God. The transcendental rationalization of experience and sometimes of religious belief undertaken by Kant and

Husserl is an onto-theological projection onto experience, because the scientific model they have at their disposal sins either through its too mathematical and transcendent character (Husserl) or through its outdated model of physics—these models, what is more, being each time confiscated by a sufficient philosophical decision. To pursue more radically the transformation of the theo-christo-logical doublet, we must abandon the old rationality, itself mixed, and invent a generic device that proceeds in a quantum manner—that is to say, through the collision rather than the mélange of theological discourses—and that is capable of undoing their sufficiency and facing up to the problem of sacrifice, the Cross, and the Resurrection. In the meantime, we will have understood Christ as the name of the duality or the unilateral complementarity of messianity and faith that transforms our theological material.

With the unformed or amorphous milieu of an absolute and purely conceptual theology that seeks to be dimensionless or without transcendental cut, like an immediate and transparent reading of the Heavens, one can obviously achieve nothing scientific: it is the mystical kingdom of fictions and phantasms, the religious and mythical ground of the Heavens. On the other hand, when theology allows itself to be structured as a system of dualities, distinguishes a lower strata, a human relation to experience, and a higher strata occupied by God, then it begins to exhibit scientific intentions; but this staging is precisely the philosophical hierarchy, and not yet a true scientific distribution of instances. This placing into duality is a fundamental preparatory gesture that does not yet lead to the positing of a matrix or a scientific vessel. A science begins by positing a more or less closed site, be it a strata or a plane, or a system for the emission and detection of the phenomena to be studied—for example, channels to capture particles. For the Christ-science we must define the emission devices and above all the detection devices, along with their parameters, like those of the Logos and the Torah. If theology is an incomplete science, it must necessarily distinguish within itself a domain or a place reserved for experimental procedures, even those of disciplinary knowledges, and produce cognizances through the collision of philosophy under the two aspects that are important for Christ, the aspects of the Greek Logos and the Jewish Torah. To produce the unknown with the known rather than the known with the unknown—such is the operation of the generic matrix. . . .

The end of sufficient theological representation is obviously not its absolute negation, nor even that of the religious, but the salvific critique of the *Principle of Sufficient Theology*, which here would also wish to be both encompassing and determining. We place theology under determining, or underdetermining, condition in-the-last-instance, using quantum theory but certainly not under positive quantum conditions. However, this is not a matter of mysticism properly speaking, but of a science that ends up not by falling absolutely into silence or, for example, into a stance of adoration, but by limiting theological sufficiency and its discourse. But if Greco-classical representation is undermined, then there is no longer any reason not to interrogate the theology that is the superior and systematic part of religion, and the christology that is its heart, as to their corpuscular way of thinking and their semantic mode of belief. Theology resists, just like the Church, like the Church-form of philosophical thought.

THE MATERIEL FORMALISM AND ITS INTERPRETATION OF CHRIST

We offer a "materiel," but not at all materialist, interpretation of the central sequence of Christianity (from the Cross to the Resurrection). Materiel indicates a certain unity of matter and a priori form, in any case, but there are many interpretations of this general proposition. Let us begin by excluding three transcendent interpretations. (1) It is not materialist or reductive, finding its sense beneath phenomena, in ethno-historical events. And yet it is a sequence of lived-materiel events, but of such a nature that we might call them wave events, interfering wave events, events that must be able to explain why the Cross is perceived as a series of corpuscular events whose sense is always elsewhere than within them, according to the traditional linguistic model of decipherment. (2) Neither is it idealist or idealizing, locating sense beneath phenomena in a sphere of spiritual significations, as if they were a metaphor or symbol of a meta-historical event or of a plan of salvation fabricated elsewhere, an object of theology that would give it its spiritual meaning. (3) Finally it does not indicate the fusion of matter and sense in an a-priori-and-transcendental identity, the identification of empirical phenomena and their ideal sense, as in Nietzsche's and Deleuze's transcendental empiricism. However, this interpretation starts to exclude brute matter and myth, and even the duality between infrastructure and superstructure; it makes their identity alternate around syntaxes of difference. This identification with various transcendental degrees of formalism and materialism remains at the heart of philosophy, and may simulate the materiel effect we seek generically, as a quantum occurrence, but precisely not in a fusional identification, or even a Moebian alternation of the two sides of one unique surface.

What is opposed to all these philosophical corpuscular solutions that go from identity to difference and to the dialectic is obviously the quantum act of vectoriell superposition, which responds to a logic other than that of dualities such as form and matter, all of which are inscribed in the doublet of transcendence such as it is reflected both in form and in matter. Superposition stems from a property of certain algebraic operations, it is idempotence $(A + A = A)$ that escapes an analytic or even synthetic operation as classically defined. The dualities of instances, originally of the corpuscular, classically nonsuperposable type, are found under certain conditions in unilateral complementarities. There is superposition when immanence is the same through and through, and traverses the transcendences that it brings about rather than contains, but that do not change it by being added to it. Two terms can be superposed if they are of such a nature that they conserve the same unchanged immanence despite their adjunction or their addition, which reenters under or in immanence. Not just any matter whatsoever is susceptible to superposition, it must be of a wave type; and not just any logic works for it, it must be an idempotent logic, one that responds to the imaginary or complex number.

If, for example, two interpretations are classically and tendentially possible, Christ-according-to-the-Logos and Christ-according-to-the-Torah, there result two types of solution that in each case involve a certain conception of matter and a certain conception of syntax. Or else they are mixed in various proportions, but the resulting interpretation is therefore always of a philosophical, corpuscular, or macroscopic type, and the macroscopic entities form mixtures or amphibolies but do not superpose onto one another—this is the state of most theologies, even Trinitarian ones. Or else, without mixing or even combining, they are superposed and we must then suppose that they change their theoretical and physical status, and must be thought fully as quantum phenomena, at once wave and particle (no longer corpuscular). It will be objected that the leap from classical matter to quantum materiality is inadmissible, a metaphorical passage; but this is to forget that the atomistic and Newtonian "natural," which philosophy largely adopts, is no more natural than quantum matter, and that it is precisely the latter that we can learn from, not the Newtonian version. We must therefore assume from the very beginning a preparatory theoretical and scientific "conversion," a quantum-theoretical stance and not a philosophical position, admitting that theological concepts, utterances, and axioms are not "natural," that their macrophysics is one possible state but is not normative nor the only state. The possible interpretations of Christ are those of superposable wave (and particle) phenomena, considered from the quantum-theoretical angle, something that is only metaphorical for the philosophy that impregnates theology, but not beyond its bounds. Insofar as they are superposed in the matrix (and thus, we shall say, superposed on the Cross), they obey another "logic," an algebraic logic, one of idempotence—namely, their addition gives a unique interpretation in quantum-theoretical (and thus nontheological) terms, which "completes" both one and the other. This "completion" without materialist reduction or idealist elevation is what we call a generic reprise rather than a repetition. It signifies that the new interferent interpretation of Christ, his new "wavelength" or state vector, at once constructive and destructive, is the "Same" as each of the "incumbent" interpretations, but exceeds them from the point of view of their analytical content by forcing them to remain the same or sterilizing them from the point of view of their synthetic content. What happens in that matrix which is our unique point of view, that of the Resurrection, is the constitution of the vector of Christ as containing all possible information on its subject, that is to say, its amplitude of messianity—which leads us to say that we are assured of his indeterminate messianity, but not of his exact messianity.

We call materiel or immanental formalism this conception of the Real that allows us a new reading of the three phases of the foundational drama—the Cross, the Resurrection, and the Ascension (with Evangelization as the fourth moment). A sort of phenomenology without materialist or idealist position, but entirely immanent, flush with the events or transcendences that enrich it. These moments are thus, in reality, phases of the same process rather than dissociated stases linked together by a transcendent guiding thread that would be in the hands of God. Philosophy has made the Cross into a scene of sacrificial tragedy followed by a separate and, what is more, unintelligible conclusion; the Resurrection and Ascension are supposed to give a final but supplementary meaning added to sacrifice in the form of a marvelous or miraculous outcome. We understand that the Hegelian philosophy (but it was not the only one) grasped this determinist and already speculative interpretation as an unexpected boon, and gave birth to the great cohort of idealist or post-Hegelian christologies. From our point of view it is instead an experimental drama that must be interpreted as immanent, not as being organized and manipulated by a Great Planner. It finds a theoretical and physical model in quantum physics, but cannot be reduced to this model; nor refocused into a human or generic physics. The Ascension closes the great theological tale of creation, which must be doubled by a new scene of mediation, as Christ will be in his turn, in the echo or the repercussion of his return. The matrix we are constructing to replace this tale and to sketch out what we must call an experimental theology has to deal with the same ambivalence between a classical interpretation of its apparent and successive movement of preparation, production, and consummation and an interpretation according to a real movement in which it is obviously the Resurrection and the Ascension that are prior-to-first and explain the sacrifice of the Cross or the abasement of God. More exactly, it is so little a matter of inverting the apparent movement of history and theology into a real movement of determination in-the-last-instance by the Cross that in fact the Resurrection and the Ascension are prior-to-first and not simply first, as they may have been in a "first theology." The distinction between prior-to-priority and priority is more than a simple inversion of the movement; it is the placing under underdetermining condition of all Christian theology. The prior-to-first Resurrection and the Ascension are less miraculous, more intelligible, than the supposedly sufficient sacrifice that succeeds only in continuing the ancient barbarism of religions.

QUANTUM ORIENTATION OF THE CHRIST-SCIENCE

Different formulations of this enterprise are possible: christic science or Christ-science, or, on the shortcut model of "nonstandard theology," a non-Christian science or "nontheology." To take account of its effects on theology, "nonstandard theology" can be used, or else "experimental theology" in memory of the passion of the mystics. A multiplicity of senses replacing the multiplicity of names of Christ. It is important not to define it by way of some part of philosophy, as if it were a question of an "ontology of Christ" extracted from the philosophical mass or a "christology" extracted from theology, when it really is a science that operates this desuturing. A science-according-to-Christ is in no way a part, a region, a domain, or a stratum of theology. This is its way of desuturing itself from the historical context in which it is adopted. On the other hand, elements of the latter are "input" as symptomal material into a "theoretical machine," inscribed in an artificial space or a vessel sampled from a certain physical rather than mathematical model. Quantum theory may be—under certain restrictive conditions, those of the principles of its algebra—as operative for theological disciplines as it is for human disciplines in general. Recall that between Christianity and quantum physics, constituting the link between them rather than their still improbable "synthesis," there is a great Christian thinker who had some influence over the constitution of this science: Kierkegaard. We cannot explore this affair in detail, but it is impossible for us not to see in it a sign or a symptom in respect of the qualitative leap, the antidialectical and antilogical discontinuity, proper to quantum paradoxes. Their extension to the "crucial" experiment of Christianity, to the Cross (including the Resurrection), will seem an unlikely provocation, as will the attempt at a christic physics. In the sacrifice on the Cross, in the empty tomb, and in the Ascension, which are the laboratory of Christianity, we meanwhile find an incalculable dialectical experiment underway, one that is philosophically impossible but relatively intelligible for an entirely other rationality. It is obvious that there is another leap between the critique of Hegel's "logical" immanence in the name of the transcendence of faith and the quantum critique of transcendence in the name of a vectoriell immanence (a nonlogical leap, it is true). The inversion by quantum physics of the theoretical debate between Hegel and Kierkegaard, which remains internal to philosophical theology, takes place by means of algebra, whose property of idempotence and affinity with the imaginary number inverts the relations of immanence and transcendence.

In strict terms, in regard to this use of quantum physics as a model (in the scientific, nonphilosophical, or paradigmatic sense of the word) of the Christ-science, we may speak of a weak quantum or protoquantum theory, as have certain epistemologists in the human sciences. But more adequately, according to an expression that is not one of evaluation but one of structure and use, we may speak of a "quantum-oriented theory." It is a question of a generic quantum theory: the use of quantum theory here is nonpositive and nonmathematically sufficient, even though it requires a minimum of algebra in order to have an objectivity. It implies the human "subject" and its operativity as observer in the very form of the objectivity of knowledge, its integration into a unique state vector. No longer an objectivity of knowledge in itself, but one that operates through entanglement or nonseparability, through the unilateral complementarity of the traditional actors of science, "subject" and "object." In particular, if our use of quantum theory is placed under condition of its weakening in terms of its mathematical sufficiency, this weakening itself is not the doing of the quantum theory and of its mathematics reduced to several axioms; it is the doing of the generic conception of the quantum apparatus. The vessel into which we hurl theology demands that we also hurl into it classical quantum theory. The generic matrix such as we understand it supposes two variables—one theological, the other itself quantum—whose fusion or nonseparability must be ensured, but under quantum condition in the last instance, and not under theological condition (this may have been imaginable, but certainly would not be "imaginary," and so was out of the question here). The generic matrix is precisely that operation that removes the positivity of quantum thought and the sufficiency of spontaneous theology, and this through the twofold means of their reciprocal multiplication or translation and their underdetermination by a specifically quantum act, superposition, and a generic one, idempotence. Whence a deobjectivated objectivity, a deindividualized subjectivity—that is to say, in both cases, generic. Being matrixial, our device conserves an algebraic (imaginary and complex) kernel, stripped of numerical calculation: this explains the quantum force, which is the force of a weakening of theological sufficiency. And what is more, it conjugates this kernel with a characteristic of the lived with which it is superposed and which testifies to the human and not solely "mathematical" nature of this generic.

We sum up the end of the identifications and specular doublets of theology and christology in the concept of "christic messianity" obtained through the quantum treatment of the theo-christic doublet. This is obviously not Judaic messianism, although it owes much to it; it is a generic or human messianity, a messianity of-the-last-instance that guides from within and configures the faith of the faithful, tearing them away from beliefs activated by the world. In its ultimate essence, faith is messianic: it is an immanent praxis for a deindividualized messiah, one that destroys the world-form beliefs of theology.

FROM FOUNDATION TO GENERIC CONSTANT: THE BEING-FOR-MAN OF THE CHRIST-SCIENCE

The principal or first act of a philosophy is that of "foundation." It delimits and encloses a natural space that will be called a "foundation." Qua "natural," it is generally a projection onto the pregiven site par excellence, nature. The first act of a science is to delimit and close off an artificial space, a domain of objects circumscribed by a constant, for what will be called an experiment. To map out a pregiven site and to construct a vessel are very different acts of definition of a theoretical "site." In a sense, epistemology, like philosophy of science, is founded on a confusion between these two acts undertaken in the name of a "site." A science of theology, where the relation is apparently inverted, seems very close to this confusion—it demands at once the vessel and the constant of a science, and an already pregiven object. Thus it is both artificial and natural at the same time.

What distinguishes theology as a science practiced by the Churches or their theorists from a science of or for theology, as a "nontheology" must be? Certainly not the inversion of subject and object, since each of them is reflected in the other and each of them is both science and theology. This site at once natural and artificial, received from physics and constructed by history, this experimental vessel that founds *quasi*-theologically a new science, we call it the "matrix," for reasons pertaining to mathematics and science fiction. And we posit that humans are not in this matrix in the same way that they are "in" the world, but that the matrix is "in" humans—that there is a "being-in-generic-man" of gnosis as science of Christ or as generic-oriented science. Humans are no longer defined philosophically as beings whose vocation is to be, but as identically natural and artificial—that is to say, "generic"—*without our knowing, still, what is this obviously fundamental and enigmatic "at once" that is the object of the cognizance we seek*. A generic matrix contains apparently complex relations that escape the logics of philosophy and science, supposedly separate as knowledges or disciplines in themselves. To simplify, such a matrix obeys two or three quantum principles: it undertakes the *superposition* of science and theology, it exploits their *entanglement*, and it acknowledges their *noncommutability*. There is thus no point in locating oneself in the "sites" or topoi of classical theological discourse, except qua clarifications or references (our quilting points with historical Christianity). We will still be detected, if not displaced, in them. To sum up, it is a question of redistributing the classical relations between science and theology into other relations or nonrelations that obey a thinking that is quantum-theoretical in spirit, even if it is not a thinking of physical matter.

Why pursue such an enterprise? Because the philosophy that ultimately governs these relations is neither a sufficiently artificial nor a sufficiently natural element for what we want to do. It is not a good artificial and experimental milieu for science; it remains an arbitrary, barely even axiomatic decision. Nor is it a good natural milieu, for it flees into theological abstractions and substitutes for faith beliefs and images traced from idealized figures, it disregards the lived materiality of faith. More generally, the element of theology is an indeterminate generality, a poorly structured space where concepts go in every direction as in a divine doxa (the true content of the theological understanding) and are continually stopped for questioning and then put back in the race. It is understood that God supports this whirling narcissistic madness of philosophies, and perhaps he is the first to participate in it; but how could Christ bear to climb onto the carousel where theologians and priests hawk their beliefs—he who came to break up these mélanges and chase from the temple of the world the spirit of corruption that is its very principle? There is probably no point in opposing it with a sterile purity or, still worse, the ideal of a *universal hybridizing*, that new desirable purity of mélange. It is an entirely other solution, in a nonphilosophical spirit, that we seek: an *entanglement* that simulates this universal hybridization without being it. More precisely, generic entanglement—this is what we oppose to philosophical hybridization and, among other things, to the technological interpretation of Christ as a transcendent union of two natures. We define humans by way of properties that are neither purely natural (animal and physical) nor purely artificial (rational and cultural), not by remaking the mélanges, but instead through the three quantum-type properties that make the human being impure yet nonmixed, superposed yet nonidentical, entangled yet noncommutative.

To sum up, the inputs to the science-in-Christ are as follows. (1) On the one hand theo-christo-logy as final macroscopic object to be explained, and simultaneously (given the definition of this generic science) as a variable that enters into the cognizance of this generic science. This theological variable itself is decomposed into two subvariables that are here, for Christ, the Greek and the Jew as ingredients of christology. Thus the traditional element of the treatment of christological problems becomes a phenomenon to be explained, all the while conserving a function as a variable or parameter of the generic as testimony to its old, royal domination. (2) On the other hand, theo-christo-logy is conjugated with quantum physics itself, with its major principles, which are treated in turn as a necessary variable of the knowledge of generic humanity. Quantum physics itself is a scientific model designed to replace the Greco-philosophical, often Aristotelian rationalism that was presupposed in the old theology. The model, just like the object or the experiment, also functions as a parameter of generic idempotence, and must abandon its empirical-scientific sufficiency, just as theology must abandon the sufficiency of its spontaneous exercise.

So are there two or three variables in the problem of Christ? There could be many abstract versions of the quantum science of Christ, with Greek or Judaic variables that could subsequently be recombined. But for reasons of the limitation of our possibilities of manipulation, we examine immediately only the elementary relations of the Greek and Judaic that are to be found in the logia and utterances of Christ as a function of the phases or sequences of his mission, without dissolving them into a "plan for salvation." This is the index of our finitude as contemporary quantum observers of Christ alone, of his sayings and his acts, not of his interpreters or indeed of a supposed plan of salvation that would wholly exceed, through a transcendent religion, Christ-in-person and his generic signification. It is a refusal of the phantasm of a total and planified view that unfolds between Christ and God within the theo-christo-logical doublet.

The alternative to "pure" spiritual experience or to the discourse of knowledge constitutes the materiality of this experience. There is no vague or abstract dogma separate from language. We must therefore decompose the Jew and the Greek into specific variables and treat many among them as materials of this science. By way of anticipation, we shall make the product of these two variables each time, but in reducing their transcendence through an immanent act of superposition, as the matrix requires. This procedure is the only one that will save us from both Jews and Greeks, the only way to extract the originality of Christ-thought that one might have thought a means or a weak combination of the Greek and Jew, or at best the schematization of transcendent God within the Greek and its immanence. Messianity is something other than the schematizing synthesis of opposed idioms. Levinas, Hegel, and Kierkegaard were such attempts to surpass schematism, to crack it open through an excess of transcendence—and even Hegel. Nontheology is another solution, which associates transcendence and immanence unilaterally.

Ultimately, the device of the generic vessel properly speaking prepares the fusion of these variables through their conjugation or reciprocal translation and through their underdetermination. This new term only forms a third objective or autonomous instance for theology, which supposes it as an in-itself; but it is not an independent instance or a third disciplinary source. In other words, the new quantum-theoretical rationality, far from being imposed from without upon the christic experience, or introduced in a forced, formal, and transcendent way into the christic matter, like logic into philosophical matter, is rather a state of immanent fusion with the theological experience of Christ. This immanence that does not at all prohibit its heteronomy to the theological subject that it affects and transforms. The generic is human immanence as principle of the transformation of the world, and thus of theology. Generic humans do not form a third term that one could imagine on the model of a transcendental triangularity articulating experience and rationality.

The generic matrix cuts short the necessity for the schematism and for the synthesis by the transcendental imagination (Kant) that are required by the empirico-rationalist hiatus, or for subtraction as specular planification required by the khôra as Platonic hiatus. One of the effects of the quantum matrix is to avoid the traditional solution to the problem of Christ, in the form of a schematization of the Jew in the Greek and vice versa (as attempted by Levinas, when he translates Judaic nonontology into Greek), mixing inverse translations of the two idioms or two principal historical references. Specifically, schematization is the attempt to insert the Jewish God in his transcendence into the immanence of the Greek ontology of man and the world. We prefer to eliminate the various mixtures that compose the spectrum of Christianity as unstable system of beliefs and images, dogmas and representations. We do not think on the basis of these historical modalities, which are at best materials or symptoms. We could call this a non-Christianity, taking the problem on its broadest level, but neither historically nor speculatively. Christ does not belong to an act of the transcendental imagination; he expresses the radical genius of man rather than that of a philosopher. There is an endless list of these mechanisms of suture or of the reciprocal mediation of science and experience, of being and thought, a suture present in all the hiatuses of philosophical (not to mention theological) dualisms, in the theo-christic doublet and its innumerable triads or trinities charged with operating the analyses and syntheses necessary to these dualisms. Two fundamental generic concepts will be substituted for these aporias that set theology wandering in the philosophical labyrinth. They are the concepts of "mediate-without-mediation," which formulates the constitution of Christ as generic subject, and of determination-in-the-last-instance, which describes its function in this science, a function that can be summed up as the placing of theology and christology under underdetermining condition by Christ and his generic messianity. This science—invented rather than sought, discovered rather than hidden—is thus a science of unilateral duality on theological soil, the nonmetaphysical duality of messianity and of faith acting in the world according to a new type of complementarity, without the convertibility of the One and of Being. Ultimately "Christ" is the name of a unilateral duality, of the invention as much as the discovery of a generic constant that circumscribes the field of phenomena of discursive and spiritual belief localizable in the human sciences, but localizable under condition of a faith or a fidelity without belief, which will render them, this time, nonlocalizable. That which can be characterized, rather hastily, as a nontheological theory of theology could run the risk of a vicious circle between Christ and the science he determines; but since it is unilateral, this complementarity breaks with circular reasoning. The return of Christ, his Second Coming, then takes on an entirely different, no doubt more prosaic meaning, stripped of its mystery but more rigorous—it is simply the scientific and generic transformation of theology. It is not necessary, in the work to be done here, to distinguish the matrix as a general case (its invariants and its principles) from its particular concretization or effectuation in such and such a material. The particular or rather generic case of Christ must suffice for us because it is identical to the general case or the "ground" (as generic, not as a theologico-metaphysical abstraction). The operation of the observer included in the unique state vector that is valid for Christ and for the observer-his-contemporary is a resumption or a reprise, a new superposition of the science-in-Christ as, also, a generic science.

THE GENERIC FUSION OF CHRIST AND THEOLOGY "UNDER" CHRIST

Let us recapitulate the matrix of the christic science; it is *the fusion, in the person of Christ, of messianity with theology, but under messianity—their superposition.* We shall explore several effects of this in the domain of religious speech and the speech of faith.

In general, we place theology *under* an *underdetermining* or *determining in-the-last-instance christic condition.* This all-important nuance signifies that this condition is not affected by that which it determines, unlike what theological sufficiency postulates when it claims at least to codetermine faith by applying "understanding" or "explanation" to it. It is an extension of the quantum-theoretical principle of the noncommutativity of quantities measured from an "elementary" phenomenon.

Christ is an ultimatum for theological thought, an obstacle that cannot but impose a step backward upon it in the order of causality, sending it back to an appearance, if not an objective appearance, or to an occasional cause, but while extracting its properly christic kernel. Theology in almost all its forms, and in particular the "dogma," is the theoretical armature that holds together institutions or Churches everywhere in their communitarian sufficiency, but that cannot claim to penetrate into the essence of the messianity of faith—and this for reasons that are not themselves theological, but generic. One of our most general objectives is to no longer discuss theological positions, but to change the paradigm or style of thought (that is, of faith), no longer doing anything more than mobilizing theology as material, rather than pouring faith into its mold. We make use of theology as of a necessary aid or an occasion to speak the faith or the "knowledge" that we are without knowing it, depending on theology only in order to create the theory of this ignorance, but no more. It is necessary to substitute for the concept as "macroscopic" form of thought that disperses faith into local and/or global representations—into molar or even molecular, but always corpuscular, entities, in short into phenomena of belief—a thought that would be the lived exploration of messianity, the affect of its invisible outflowing, of its entirely straightforward rigor. It is less a question of living the interiority of individual and, in short, egological faith, of being more "pietist" or indeed more obedient, than of following its immanence as flux. Christ is formally, or through his theoretical status, an indivisible wave of the lived that is neither historical nor worldly, and that no longer accumulates in the form of a dogmatic treasury, but that has a generic universality.

Generic means first of all that the faithful lived can absolutely not be singular, nor banally "common" or representable through institutional codes, but that it is valid a minima for two individuals. But this is still a very poor formulation; the generic is not the extension, under the same principle, of an ego to an intersubjectivity. It is not even a between- or interfaith crossover, but an inclusion of the Two of belief in the immanence of faith or of the One of messianity, a faithful inclusion of my ego in a messianity. More assuredly, a faithful drive supporting itself on this Last Instance, emerging from it, and carrying itself to meet us, the world-subjects, that it traverses or suspends in a "cloud of unknowing." The generic is the unicity of a unifacial duality, one might say, a phenomenon of outflowing or up-turning by faith of that which is belief. Here the mystics are our masters in language and experiment. But it must be admitted, in order simply to make our project intelligible, that in quantum physics we find a thinking of radical but generic immanence from which we extract the "quantum-rational kernel" so as to test it out in other spheres. The notion of "rational kernel" should not surprise us, since it is a question of establishing christic messianity or faith as the principle of a science of religions.

THE GNOSTIC ALLIANCE OF SCIENCE AND THE LIVED

This type of intrinsic alliance of science and the lived necessitates the invention of some new fundamental concepts. (1) It is an immanent fusion—that is to say, a fusion *by means of an immanence, and thus by means of a mediate-without-mediation and of immanence, of an act of superposition as mediate-without-mediation*, and not a mere suture as a transcendent operation between separate and "corrected" terms by an immanence (an external or internal relation to which we oppose unilateral complementarity).

The algebraic "logic" used here is that of idempotence, of the flux in immanent excess over itself that constitutes the subject-science—what we also call the lived of messianity. We thus interpret the immanence of faith and of thought "in-Christ" on the basis of our two principles that form a quantum-oriented thought but that we could also call "gnostic." It is not exactly that we wish to give historical gnosis the status of a logic, but we identify, if not in logic then at least in algebra, a gnostic operation, idempotence, as used in quantum physics (but irreducible to the latter). Gnosis is traditionally an unacceptable doctrine in philosophical and especially religious matters. But we shall see that gnosis here is scientific, and brings into play a lived materiality—we do not say a materialism; it thus refuses the rationalist, positivist, or dogmatic condemnations that, so as not to lack sufficiency, end up simply lacking plasticity and freedom.

We could conjoin or suture a positive hard science and the subject—for example, mathematics and the religious subject—but then there would be no possible science of religions, just a sort of vague and dogmatic confirmation of Christianity through mathematics, just as the latter already indirectly confirms philosophy. On the other hand, the insertion of the subject into the quantum principle is possible and, in truth, necessary. Quantum science makes *possible* or plausible a nonpositive concept of science as the alliance of scientific procedures and of a generic subjectivity that inhabits these same procedures. They do not suppress the latter, but transform it by obliging it to distribute itself across the whole space of its procedures. Quantum physics as rational body of work can then serve as a scientific model for the Christ-advent. Again the two principles invoked must be detachable from their original discipline and transposable into a subjective science, a science without dialectic and without individual ego, both of which would bring it back into the lap of the world or of philosophy.

CONTEMPORARY OF CHRIST THROUGH GENERIC SCIENCE

Since it is a question of the doublet, let us straightaway resolve in principle the problem: In what way are we contemporaries of Christ?—a problem we shall rediscover later. Contemporaries of what or whom exactly: of

Christianity, of Greek philosophy, of Judaism, of their histories or their institutions? Is Christ himself their contemporary, as was Jesus? We are not the contemporaries of these ancient and modern knowledges, but too often the victims of it. The fact that Christ is not their contemporary but their salvation allows us to be his contemporary, and his alone.

The point of view of Christ-in-person as generic being and of our being-contemporary-with-him allows us to liquidate the mythological representations of Creation and Incarnation, and in general the aporias linked to the necessity of a schematization of the theo-christo-logical doublet. We insist that here we find distinct problems with their own solutions. We must not confuse the variety of expressions and conditions of existence of the problems with their general form and their invariants, even if philosophy is accustomed to do so, precisely according to the redundant logic of the doublet. The problem of the man-God relation must be settled otherwise than through religions and the vagueness of the theological terms of Creation, the philosophical terms of Schematization, and sometimes the physical terms of Incarnation. These are often borrowed clothes: theology is a garment that is too vast, too transcendent for the generic Christ. His being-contemporary with himself and henceforth with us too, or his being-superposed with himself, and therefore with us too, is the sole measure of our common and future incarnation. We are contemporary through his messianity, not through a supposedly common historical past or present.

THE QUANTUM OF FAITH AND THE OCCIDENTAL-ORIENTATION OF THE SCIENCE-IN-CHRIST

The "Christ" event is in each of us, we harassed believers of the world. So the generic fusion of science and subject is involved in a messianic undergoing. Each of the faithful in-the-last-instance is under-going in suspense the weak ultimatum that can be opposed to History, the Good News that is practiced prior-to-priority of every ecclesial and dogmatic elaboration, and that underdetermines them. Christ is not the founder of a religious tradition except insofar as he himself is, prior-to-priority, the name of a lived science made of oraxioms that will come to invest this Greco-Judaic tradition and transform it into an occasional material.

We no longer confuse faith with a belief instrumentalized by the Churches. Belief is representation or its equivalent in religion; it is ecclesio-molar or sectarian-molecular. So the confusion of faith with belief is of the same type as the confusion of invisible vectoriell immanence with the corpuscular object. From the point of view of philosophy, faith, with neither a macroscopic object nor a subject, is perceived as an apparent abstraction of belief. But belief is ultimately what destroys faith in its superpositional essence, and the Messiah himself.

The generic messiah is not a telos of the All of world-history, but a force for its transformation. He functions like an algebraic-style idempotent addition. Messianity introduces a perspective that is a wave rather than a corpuscular perspective; but it is very much a question of tying together, independent of the combination of a Far Eastern–style quantum thinking, the alliance of a Christian-Western quantum science that is equally opposed to a philosophical marginality that is a corpuscular falsification of the messianic and that must be overthrown and displaced. It is the means to establish not a quantum Christianity but a "non-Christian" quantum science, we could say, limited to theology's Western horizon, quite unlike its Far Eastern absolute and unlimited "wisdom." We do not confuse philosophy, which is certainly an infinite form, but one delimited in itself and upon which a science can therefore be grafted, with that vaguely outlined wisdom, with margins and lines of flight, too close to a doxa. Philosophy and the theology that crowns it—and this is what distinguishes them from this wisdom—are the form of the world, which itself can be treated as an object of science; this is precisely the first act of limitation and determination demanded by a non-Christian science.

Physicists have already "tied together" quantum science and the Tao as human wisdom, but without operating a "quantum science of the Tao," which would be our true scientific project, rather than a "Tao of quantum physics." So it also cannot be a question of a Western and Christian repetition of the Tao of physics as if through the superposition, practiced since Schrödinger, of the wisdom of a Far Eastern religion like the Tao and quantum physics, a superposition under the tutelage of the "All." Our attempt may have a similar allure, but two enormous differences distinguish it. On the one hand, the observed phenomenon is the Christ-event, and not some already-constituted religious wisdom. The discourse with which quantum science is superposed or to which it is added is Western theological discourse, a religious philosophy whose outlines are relatively well defined. The difference is between a radical determinate science of the All (even when it recognizes its imaginary unlimitedness) and an absolute thought of the All, the philosophical, which works in the All as its intraterritorial element—that is to say, nowhere—and thus repeats its initial presupposition. The quantum of faith or Christ-quantum is the necessary condition for delimiting these religious discourses and showing their limits, or those of Judeo-Greek theology. And on the other hand (a second difference), we manipulate the superposition and the state vector as a function of a certain philosophical precision. This is a necessary prudence if we wish to distinguish firmly between the philosophical and the generic. We call quantum of faith the Christ-event, the inevitable minimum of new faith, the minimal definition of thought-as-faith that no longer relates as belief, sentiment, or concept to an external object or event. It is the constant that science must observe in its examination of humans as a faithful condemned to belief and bound by the world.

WHAT IS A GENERIC SCIENCE?

The sayings of Christ express a stance that, from an external point of view—to the categories of historians and exegetes, for example—seems interpretable alternately upon the basis of either Greek or Judaic notions. But it is the whole of this *oscillation* that must be taken as a nonseparable albeit unthinkable whole, as an indivisible wave of speech whose secret we perpetually seek. It testifies to a stance that is *unilaterally* indifferent to the objects that his sayings bring into play yet that, on the other hand, are the corpuscular obsession of belief. His force, which certain mystics have experienced as a sort of experiment, consists in fusing a practical rigor of thought with a faithful or messianic lived.

What is important in the sayings of Christ, and what is irreducible to the letter of the logia of Jesus and to the positive criteria that allow us to collect and compare them, is their at once immanent and messianic emission, which is no longer that of a prophet relaying a divine word, but the vectoriell and physical body of this word that brings with it the seal or the mark of its authenticity. Messianity is the index that distinguishes faith from belief, at the same time rendering them nonseparable in a different way each time. It is an immanent indexation of the logia as variables to the "logic" of their superposition; it puts an end to the transcendent age of the Jewish prophets by heralding the age of the faithful, beyond that of the Apostles.

It is through the vectoriell reduction of these variables taken from Jesus's logia that our stance can be called generic, distanced from any theological position. We define the generic through a certain type of conjugation of science and of the lived of belief, a superposition that creates another state called faith or fidelity. This new state, a state of affect as much as of thought, is not a third state alongside the two others—as if one could count them and, for example, imagine a minor form of the triad that can be seen to grow macroscopically, like a Trinity. The age of fidelity makes it impossible to enumerate divine persons or to theologically calculate salvation. Not only theologians but even God finds it impossible to plan out a plan of salvation.

The generic stance is the stance of humans no longer believing but finally knowing their faith or their nonreligious messianity. The genius of Christ, if not of Christianity, is to have revealed the generic essence of sciences that can only claim to be "human" but that do not engage the lived in a radically immanent manner, in their stance and not only in their object. The lived has nothing egological about it, it is part of the protocols that prepare the science-in-Christ, it paves the way for the procedure of idempotence that, for its part, neutralizes it generically or suspends its philosophical characteristics. There is a lived that is neither an All like life nor a singularity like the living creature, but that adheres to a scientific procedure of idempotence—it is generic faith. The theoretical novelty lies in this generic quantum-type fusion of science and the lived that is the sign of the discovery not of a new soil, terrain, or foundation, but of a stance of-last-instance. To say it more simply, it is *a new practice of fidelity as "in-the-last-instance."* And the discovery is that there is not a subject of science, in the modern manner, whether that of Descartes or of Lacan, as if science were already the predicate of a preexisting subject that would be Christ, nor even a post-Lacanian subject transited by the materialist void, but only a science "for" the subject (*science «à» sujet*) or a subjective science, where it is Christ who is, to say it too quickly, the predicate of this "subject"—all terms that we doubt can be transformed, their logic becoming immanent. In particular, strictly speaking, we must admit, if science is to combat the fetishism of belief, that "Christ" as Son of Man is the name that each time supports a variable in the generic matrix of science. That is to say that the new alliance can be conceived neither in the Greek manner as a heroic alliance of man and Logos in the form of philosophy nor in the Judaic manner as the exceptional alliance of God and his people, but only as generic, as an alliance of science and belief conjugated *under* science, that is to say, as faith or fidelity. Ultimately, if philosophy takes into care beings as a whole, God included, the generic takes its whole being, God and man included, as a function of man, but this time as generic.

THE SCIENCE OF RELIGIONS AS THE DEPOTENTIALIZATION OR DEBASING OF BELIEFS

We know that quantum physics, being a mathematical "formalism" capable of predicting indeterminate events, is inseparable from another whole theory of its interpretation in terms of reality (realist, idealist, pragmatic, and so on). From our point of view it is less "positive" than classical mathematics and physics, for it already implies a subject—only a physical subject, certainly—as observer, though one who prepares it to be a science not only of particles as natural phenomena but also of the entities of classical physics that it proposes to "cognize." Just as we treat the Christ-event as that of a science of religions, but are obliged to begin from these religions as symptoms, since the religious subject is this time included in principle in the objective apparatus, as if the operation of measurement were an integral part of the objects of this science. Whence a hermeneutic appearance of the "reading" and "interpretation" of the meaning of religions that will force us to enter into our science and to disrupt its axiomatic. Given the inclusion of the observer subject, a generic science is submitted to the constraints of a nonpositive objectivity yet more difficult than that of quantum physics. Anyhow, a science of religions cannot help but also be a critique of the illusions of the theological, not only of the religious. Yet it submits the critique of appearances of the object to their knowledge: this is not the all-critique or criticism; it makes of the knowledge obtained a transformation of its object, something that was already obviously the case for Marxism. We fight against religions with the weapon of atheism as an external accomplice. Atheism and materialism are rather trivial philosophical positions without any great importance: they form an integral part of religious wars or wars "for" religion.

Neither do we have the right to an all-interpretation in regard to Christ. This is the abusive right claimed by theologians and philosophers, who certainly fix certain limits to their arbitrary will, but only within their "principles" or within the limits of their "axioms." Another right to interpretation is demanded and governed by the materiel formalism of faith, between the generic apparatus with its axioms and the chosen religious material, which is not without a reverberation in the axioms of this formalization of Christ. The believing subject of diverse religious or confessional extraction cannot but make his own accent heard, despite his becoming-generic. In this tract will be heard, for example, many reformed or still gnostic nuances—for some readers, Judaic nuances; above all, the accent is placed on the struggle against institutional forms of religion. But the Christ-science remains a weapon for the weakening of the harassment of all religions, all deadly to various degrees.

Four

CONSTRUCTION AND FUNCTIONING OF THE CHRISTIC MATRIX

THE CHRISTIC MATRIX AND ITS COMPOSITION

We shall call "christic matrix" the quantum-and-generic device that represents in vector form the possible "states" of Christ (called "Christ-in-person" to distinguish him from Jesus) and determines him in terms of a new economy of salvation, delivered from the theo-christo-logical doublet, emancipating in subjects messianity and faith. This matrix is a theoretical apparatus that goes beyond the persona of Jesus and his history, and functions as a theoretical, experimental, and lived vessel. It contains different elements represented vectorielly, among the best known of which are: (1) Materials: the data of his words and acts, with their accents of the Greek and Judaic, which are the two variables of his message. Theology and history are limited to organizing in various ways, but always within the theo-christo-logical horizon, these Greco-Judaic relations to which Jesus is reduced and to which they wish to reduce Christ. (2) The operations of conjugation, multiplication, and addition, carried out upon these variables, and that are necessary from the moment when Christ appears as a vessel or a unit of vectoriell procedures, relatively resolved or complete, but not closed or foreclosed. (3) Finally, and principally, going by the name of Messiah-factor, a special factor of underdetermination for the products of these variables, or in any case an index to which to relate them. This is a productive force of fiction or of a controlled imaginary, a force of rigorous insurrection in theology, at least once the latter is understood in its proper theoretical complexity as philosophy. This factor has certain affinities with (for example) the Real as "impossible" (Lacan) but it does not surround the signifier or the symbolic; instead, it is the symbolic that would surround it, as its margin of language or of world. The messianic-factor is the impossible for theological sufficiency (PST), but an impossible that, even so, algebra allows us ultimately to make possible, while also rendering it capable of transforming theo-christo-logical representation. The matrix uses means taken from quantum theory and from theology, but it is not a simple exporting or transfer of algebraic technique or of theological conceptuality. Certainly what is specific about it, and what will distinguish it from old uses of the concept, are this algebra and this geometry of the imaginary that it borrows from quantum physics along with its most general broad principles: superposition, noncommutativity, indetermination, and wave/particle complementarity. It is obviously not a question of a theology of quantum physics, still less a Far Eastern one. At most one could say that the complex imaginary number or, geometrically speaking, the quarter-turn find a functional equivalent in the messianic function, and that, inversely, the force of Christ's insurrection in history and philosophy can be intuitively figured by the algebraic impossible.

As to the functioning of the matrix, the messianic-factor is in general the factor of underdetermination. Its generic aspect is that of its orientation toward humans, the function of the matrix being to produce faithful cognizance—that is to say, to produce faith that is a rigorous (which does not mean exact or determinate) cognizance of messianity. If we wish to think messianity a little more rigorously as the christic function *for we who are contemporaries of our humanity* and, in order to do so, substitute a contemporary science for the old purely conceptual and hyperspeculative theological science, there is hardly any other solution than this recourse to a quantum modelization of theology. It will be objected that faith cannot be modelized by physics, but only by a more "spiritual" transcendent discipline. However, physics reduces faith into the immanence of the human lived, whereas theology dissolves it into phantasms and beliefs that develop in every lived milieu as infinite and limitless.

AGAINST THE PRINCIPLE OF SUFFICIENT THEOLOGY

The task is now to conceive this materiel (that is to say, lived) machine, capable of underdetermining or of determining in-the-last-instance theo-christo-logical difference, to place it under a complex condition that associates the quantum-theoretical model of thought with theology itself, without negating them, denying only their respective sufficiency in the name of man, whose content, in messianity and fidelity, was revealed by Christ.

To bring down the Principle of Sufficient Theology, it is no longer the theology of the Logos that must double itself, but quantum science itself, reduced to its generic usage. Faith has always been understood as a combination of Greek reason or science and christic experience: a belief. But this combination, generally speaking, must be generic and human-oriented. It must be determined at once which ultimate variables are necessary, and under what matrixial form. These latter components, as we have said, are the Jew and the Greek, but for us they are only variables within a matrix that is more important than them, and that conjugates them, the expected result being a faith that is no longer a mélange like belief, but the cognizance of Christ. These two components have always been recognized, but only as they have been mixed at the random behest of history and of theological variations, or in any case under the controlling domination of the Logos. Radicality obliges us to decompose the Logos itself into two aspects, science and philosophy. And to introduce quantum science as yet another term between the Jew and the Greek, which are themselves the ultimate components of theology as science. For this theoretical mass traditionally claimed as "science," we substitute in theology its ultimate variables, in science a new concept (quantum thought), and in the general structure of combination another more generic type of conjugation.

Four changes must therefore be introduced into theological science to meet the criteria of nontheology. Given that philosophy is already a mixture of science and logos, and that it repeats or returns, in fact the generic matrix is analyzed not into three functions but into three or four. At this level of generality, it is a matter of very general categories or symbols. It goes without saying that neither Jewish thinkers nor philosophers, each attached to their particularity, would recognize their object, concern, or passion here, and would refuse this level of formalization. But it is important to grasp that the matrix puts into play minimal variables drawn from "Philosophy" and "Science" that must be purified of their (for example) theological mixity, through which they already mutually interpret each other or anticipate each other reflexively, leading straight into a vicious circle.

1. New variables to conjugate, the most fundamental ones given our object: Christ as "system" capable of passing through several states. On the one hand, Judaism as radical transcendence or Torah, the most hyperbolic type of which is given symbolically by Levinas. And then the Greek variable of the Logos, whose most hyperbolic type is given symbolically by a series of philosophers—Plato, Descartes, Kant, Hegel—who combine immanence and transcendence according to various relations, but in reality double transcendence in immanence, or submit immanence to transcendence. In its origins, the Logos is in fact a mixture and not a pure variable; it is already a mixture of (at first premature or anticipated, then Galilean and Newtonian) science and the philosophical decision necessary to theology; it is the rationality proper to classical theological science. But here as in theology the Logos is decomposed into an aspect of dominant philosophy as variable, and an aspect of subordinated science, which represents one of the only possible scientific models possible in this globally Greek context.

2. As to the science (rather than philosophy) side of things, it is hidden, qua telos or quest, in the Logos, and must be extracted from it, replaced by X, and inserted as such into the generic matrix as a variable. This function of a scientific model = X is fulfilled by a new, physico-quantum model, replacing Greek rationality, which was under massive philosophical dominance (and doubtless for this reason fascinated by its double, if not its doublet—in Plato, mathematics). Whereas the new model allows for a radical critique of the macroscopic essence of the concept and of philosophy.

3. Another type of combination is necessary, matrixial, and generic rather than dialectical and philosophical. The variables could have been combined by a classical science and its rationality. Now, in theology as in classical science, science is required twice over, as we have just seen in

the mixture of the Logos as already-mixed science and philosophy. The doubling of science takes place under the impulsion of and controlled by philosophy, but *the generic matrix demands that we take this doubling itself outside of philosophy and conceive of it as scientific. Therefore it demands that we find a form of doubling that is proper to a science: this will be the idempotent reprise proper to quantum superposition, as opposed to a philosophical mode of repetition.* Vectoriell superposition is the weakest "repetition," the most "sterile" in a sense, as opposed to the profligacy of theological concepts; it is immanent and is invested in the conceptual-religious or in philosophical transcendence, which is itself doubled, is sufficiency. This simplification of science conserves it as an act of superposition affecting the variables, and delivers it from philosophical repetition, which doubles through transcendence. The genius of superposition consists in its simulating repetition without any recourse to the specular doublet; its principle is idempotence, which is neither identity nor difference nor their combination. And in the same gesture, we have acquired the response to the unknown X—it is quantum mechanics—and its double intervention as conjugation of variables and their superposition.

This device allows for an understanding of the possibility of a generic or faithful cognizance of Christ, the latter being what there is of the real in that cognizance that is faith. We have said that under the name of Christ it is a question *at once but in-the-last-instance* of faith as a real object, and of the cognizance (itself faithful) of that faith, or of the faith-in-person of the new faithful. The generic lived or the "Christ" whose cognizance we seek under the name of faith is itself a "function" of two variables: idempotence as the objectivity required by quantum thought, and the philosophical subject as occasional variable, this time across its entire generic form. The double intervention of quantum thought signifies that the Christ-subject is a function of a double variable, but that the "bound" variable is itself a function of the free variable (which is theological, philosophical, or "religious") and at the same time identical to the function itself. In other words, quantum thought is twice in the state of a variable, which is the work of superposition.

The general properties of vectors are addition and multiplication. Their linear addition or "superposition" is fundamental for us: it is their capacity to yield, through simple addition of the two that they are, a new vector or a new unique wave phenomenon. Meanwhile, the amplitudes may be multiplied by any number whatsoever. Ultimately, since theological vectors will here be not simply positive-quantum but also generic or said of the human Real, we modify the terminology of our categories: we speak of the "vectoriell" and of "vectoriellity" rather than of the geometric vectorial. So the new category used to describe nontheology, the name of Christ, and its properties such as messianity and faith will be that of vectoriellity.

UNILATERALITY AND MESSIANITY

In a certain way, unilateral complementarity is what, in theology, replaces the "union of two natures"—at least, one will be tempted to interpret it thus. But once this (rather philosophically trivial) generality is admitted, everything changes. The duality of the "two natures" is but an arithmetical duality relating to ontology and the transcendental. What traverses and unifies these two natures, whatever they may otherwise be, is a lived immanence whose vector traverses itself as if through a "tunneling" effect, and which, for this reason, we may call radical.

Traversing itself in this way without taking on any defined trajectory in philosophical space, it thus also traverses everything that believes or could believe itself able to exceed it, and that is at first of another nature—the macroscopic transcendence of theology. Messianic immanence is not a point or a circle internal to itself, but a radical superposition that makes an immanent tunnel, traverses transcendences thrown in its path, going beyond them without skirting around them as Being skirts around beings, as its horizon passes behind and not simply through the thing.

There are forms of unilaterality in all philosophies, but they are ultimately condemned to reversibility. We need a quantum and generic context in order to escape this. When transcendence is double, it exceeds itself as one exceeds a thing rather than traversing it. This excess of and in belief therefore believes itself able to exceed the wave of immanence; but this wave is "unsurpassable" or unlocalizable, as is the immanence of every site. The theoretical appearance of theology is to think this duality on the basis of each of its terms supposedly isolated from the other, as in-itselfs—that is to say, in corpuscular manner. In reality the "syntax" of unilaterality is neither synthetic nor analytic, because the "two" instances at issue are nonseparable on the side of "holistic" immanence but separable or local on the unilateral side of the transcendence of the world. Messianity is a flux that ceaselessly penetrates the world that tries to dam it up but that it swathes, as the night sky swathes its stars.

The messianic vector, in a sense, does not "surpass" according to the universal model of transcendence operative in all nineteenth-century philosophies, a model it does not share. With the Resurrection and Ascension of Christ, we shall see that the immanent wave of messianity ascends or "insurrects," but without transcending toward a being. Messianity is a vectoriellity with only one face. Phenomenologists would say that transcendence is "in" immanence, but in reality it is radical immanence that configures under vectoriell condition, or vectoriellizes, the received transcendence of the world. Christ's Ascension contains one possible reduction of phenomenology in its Greco-Judaic version, and partially in its "material" version (Michel Henry's transcendental philosophy of Life and incarnation). Messianity does not transcend ontologically toward God, except in theological belief.

The categories that issue from rationalism must therefore be subjected to a quantum deconstruction. To resume immanence rather than to acquire it a priori, an addition (not arithmetic but algebraic) is necessary. This addition of immanence to itself acts as a nonarithmetical subtraction of transcendence from its own doublet, which ends up adjoined to immanence. Such a retroactive transfer or seesaw movement between the "two natures," which however do not communicate and exchange nothing, is only possible because immanence unifies itself tunnel-wise with that which is adjoined as coming from the world.

Continuity and nonseparability, locality and separability, make up unilateral complementarity. It will be said that this is to presuppose the problem resolved in the very positing of its conditions. But theology also supposes the problem resolved, in an entirely other manner that in reality makes this "solution" impossible and makes a vicious circle of it: it gives itself the conclusion in the premises, or gives the premises in the anticipated form of the conclusion to be obtained, and can neither generate it nor do the necessary work, since it contemplates the conclusion as solution. Strictly speaking, then, what is exchanged in the classical question of the "two natures" is unilaterality itself, or the generic constant that passes from man to the world, now investing the latter as water invests sand "from within" and comes to the surface. But nothing is exchanged in the interior of unilaterality, since the duality of the "two natures" is not an arithmetical transcendental duality or a modified ontological duality. Unilaterality is in no way a transcendental, still less a psychological, interiority.

"CHRIST" AS QUANTUM VECTORIELLITY

In accordance with our problematic, which may appear scandalous (not religiously but theoretically), we cannot take into account extreme protestations, whether Christian or materialist, and we will admit that "Christ" is the name of a lived wave that is materiel (not exactly material) and/or of an indivisible particle, message, or kerygma that has the force of the "imaginary" irruption of a "complex" entity whose indivisibility renders it philosophically unanalyzable, theologically indecipherable. To say that Christ is the name of a particle borne by a vector is not to make Christ a brute material force—on the contrary, insofar as, qua particle, Christ is also and right away a flux of messianity that guides or configures the particular character of its message. The christic science is far from being a sort of transcendental or dialectical synthesis that brings together conceptual contraries; it is a transformation of the latter through quantum means, means that anticipate their fusion or their superposition.

What we call the "state" of a concept, for example, a theological concept, is a set of general and particular properties. General: a concept is a system that possesses a linguistic being or entity, and the function of transmitting or communicating a signification. Particulars: the position in the space of philosophical relations and its becoming, thus a variable positional and relational being presenting, in this case, Greek and Judaic traits in various mixtures. These simple determinations are valid for classical theological entities, whose space is defined by the coordinates of God and Christ traced from the philosophical, give or take a few variations or reorganizations. But the "state" of a concept, in the complete and quantum sense, inhabits a space that is distinct from that of classical theological representation. It is a space of vectoriell (rather than simply geometrical or vectorial) configuration, because the state is endowed with a factor that changes its nature and abstracts it from theological representation. We understand Christ, in a broad sense, as message or kerygma, but in the narrowest sense as the factor that introduces into the matrix the "imaginary" or "complex" dimension of fiction. To schematize Christ intuitively without returning to the spiritual, transcendental, or theological schematism, we represent him geometrically as the "quarter-turn" of the circle algebraically denoted $\sqrt{-1}$. A schematization that we shall, of course, discover once again in relation to the Cross and the Resurrection. The state of a concept, this set of its properties, is representable through a vector in what is called "Hilbert space," a vector that characterizes the amplitude of a wave or an interferent phenomenon (the lived of messianity)—not a corpuscular phenomena, which has a position and relations to other corpuscles but no amplitude. The force of excess of such an imaginary factor, "Christ," will henceforth accompany all of our statements of a theological nature. The sense of the apparently clear fabric of signifiers and signifieds is taken up and transformed by this imaginary factor, which renders possible their wave function or vector form. Christ is the negative quarter-turn subtracted from the theological circle that, however, he generates. In other words, Christ could be written One-in-One, on condition that this is schematized by $\sqrt{-1}$. As for the evangelical sayings of Christ in the "expanded" sense of kerygma, they are what we call oraxioms, wave functions of a message made of particles animated by a vectoriell force that relates the message to a "nontheological" space; they have no meaning in the typically accepted linguistic space in which we receive them as popular language, or indeed as "concepts."

THE STATE VECTOR OF CHRIST AND ITS CONSTITUTION

Christ is the name of a particle of message or of a kerygma with no determinate trajectory (except in a presupposed determinist context), whose messianity is consequently a wave function or a state vector, containing the probability amplitude of the message being received. Let us make this formula more explicit.

Messianity acts by way of the most apparent messages of Christ that are the variables of signifier and signified, of meaning, in general; it is the state of Christ as system, the set of linguistic and theological determinations of the message. Except for a classical theological understanding that confuses cognizance with the conditions of cognizance or the object of cognizance with the real object, these conditions do not at all suffice (contrary to what is supposed by theologically sufficient belief) to give the "probability" amplitude that this message will pass through, will really arrive with humans, or that messianity will be transmissible in the form of faith.

As well as these general determinations of all discourse, we must suppose that this message is not empirico-linguistic, that it is endowed with a vectoriell (and no longer vectorial) dimension, an algebraic dimension of the "imaginary" factor that is precisely the christic factor as radically nontheological. This imaginary dimension, as we have said, can be schematized by a quarter-circle; the latter is the "negative" genetic or messianic element (algebraically denoted $\sqrt{-1}$) of the total circle formed by Christian theo-christo-logic.

One is tempted to say, following the metaphysical and classical model of the excess of transcendence as discussed by Levinas and the philosophers, that the "imaginary" or messianic Christ "pierces" the conceptual and representative closure of his message. But this is not the case: Christ is not essentially a piercing, which would suppose him to be conditioned after all by the existence of conditions of closure. The christo-algebraic condition is completed, certainly, as a new, abbreviated form of transcendence. But its positive operation is that of a simple ascending, an ascending without redoubling, an insurrection that is unifacial, endowed with a face in which it does not reflect or mirror itself. Instead, the insurrection "abases" the pride of theology and ecclesio-centrism as instituting signifiers and guardians of faith. This dimension only makes sense, only has a site, in a nonecstatic vectoriell space—this is not a thematic or semantic difference but a difference in the *reality condition* of the christic message qua incomprehensible to the disciples who follow it blindly before having the revelation of its meaning, or more exactly of its reality, at the moment of the Resurrection, and then, in the Pentecost, of the probability of its effective presence or messianity. This dimension is necessarily the dimension of an excess, since it has taken hold over ecclesio-centrism, but of an excess that is weak and consequently all the more incomprehensible or scandalous. It is the specific, nonpsychological imaginary of messianity, which is fulfilled in an operation of the superposition of acts of faith as so many vectors added linearly. The simple and messianic ascending of vectoriell superposition that, as a repercussion, brings down the excess of the double transcendence of belief and reduces it to faith as noematic immanent object. The vectorial or geometrical conditions pass into a vectoriell or generic state with the superposition of algebraic or christic "imaginary" conditions and of the human lived. The vectoriell marks not the death of God but the end of the PST, as its geometrico-philosophical double transcendence is "abased" in the christic experiment. As for the experience of the Cross, it is perhaps not without its effects upon the *usage* of mathematics.

This state of the "Christ" system constituted by the fundamental variables of the message, augmented with a vectoriell dimension, contains all possible information on the Christ-system. Every phrase or utterance of the message must be treated as a wave function, an assemblage of variables, and a complex dimension that makes of faith the amplitude of messianity, the amplitude of knowledge flush with language, in no way imaginary in the psychological sense. This amplitude of messianity contains the essential of what we can know here in this experience of Christ.

FROM MESSIANITY TO FAITH

If the insurrection of Christ as "resurrected" deconstructs belief in its duplicity or doublet of theo-christo-logical representation, a clarification of the great subjective categories linked to vectoriellity has become necessary. In the christic matrix that treats of the states of the Christ-system, materiality is constituted of two principal variables or properties, one of scientific origin, the other of philosophical or macrotheological origin. But one of them, as we have seen, must also serve as a calibrating index for the products of these variables. Whence, for each category, two opposed orientations. We shall distinguish four categories of vectoriellity: messianity, faith, fidelity, and belief.

Messianity is first of all the set of data or variables that form the Christ-system, which as we know contains superposed states. But the Christ-system was prepared in view of the experiment of faith. Reduced to the Christ-system, messianity is the condition of its cognizance as faith in the subject, but is not yet that cognizance as cognizance of messianity or result of the process of cognizance; messianity is only the probability amplitude of faith. In a more complete sense, messianity, which replaces a "divine plan of salvation," is the prior-to-first grace that underdetermines the becoming of this system, including the individual subject, which latter is called faithful or unfaithful depending on the reception of faith, and which assures (or not) the reprise of messianity.

As fusion or interference of objectivity or algebraic "productivity" and the human lived, it constitutes the materiel substance of messianity, but a materiel that has already (retroactively) been the object of a reprise at the moment, one time each time, of the adjunction of a new phenomenon to integrate as vector into messianity, and that will resume it. This reprise is the christic praxis properly so called, or the praxis of the last instance. It is irreducible to the acts listed by the Gospels and collected as the "life of Jesus," which are only the means and the occasions that messianity makes use of. Radical immanence via superposition of messianity with itself makes of the latter the Last Instance or the Christ-ultimatum required for action; its origin or immobile motor is the vectoriell quarter opening that alone permits the constitution of a vector of messianity, and subsequently of a noematic faith oriented toward the world as transcendence. Generic messianity in a sense has no object, or has the faithful subject as its primary object.

Two opposite orientations of the reprise are possible. One of them is made under the determining, or rather overdetermining, condition of the lived in its philosophical and thus transcendent provenance (spiritual, or even mystical, affects). It is a religious reprise that produces belief, and diverts messianity from its destination by wanting to impose a destination upon it, encysting it in the form of a Church or a transcendent ecclesiological body. The other resumes materiality, this time under determining, or rather underdetermining, algebraic condition; it is the reprise that we have called "gnostic" in a sense widened beyond its historical contours.

We distinguish the generic messianity that we posit as a set of conditions preparatory to the experience of faith and faith as "individual" experience determined in-the-last-instance by messianity. Faith is the individual assumption of generic messianity; it is the correlate, or rather the effect, as the reflection of messianity, and takes individuals as its supports without being conflated with them. Like messianity, faith defines the real object whose cognizance we seek; but it is not itself this object. The essence of faith is also the immanence of an ascending or of a radical but not absolute insurrection. The person of Christ named "Jesus" must also be treated as one of the faithful equal to the others, for we know that he was far from being exempt from all weakness in the assumption of his mission. As individual subject of the Christ-message, Jesus himself is also indexed to the Christ-factor.

Faith, on the other hand, is the probability of the presence of messianity in the individual subject, or cognizance of messianity. Faith is the form, the frequency, and the intensity with which messianity will get through to the individual subject, who is a macroscopic subject. This subject, the future faithful or unfaithful, is included in the unique vector that "resumes" *the Christ-system + the faithful*. As probability of the presence to individuals of messianity, faith is obtained by multiplying by itself the resultant of vectors; thus the purest or most radical faith is obtained by multiplying the vector of Christ by itself—it is Christ to the second power, and thus the sum of messianic reprises.

Faith, as knowledge of messianity, is not an individual ego, but the restriction of generic messianity to its reception by individuals. Despite the close relations they maintain, we shall not conflate faith as particle of messianity and the individual who is the form of the matter of the world and who is transited by this particle of faith. The individual assumes this particle, is seized by it, and thus becomes (or does not) a Christ or one of the faithful. The individual is a being of the macroscopic world, he is seized by messianity and becomes one of the faithful rather than a believer or, on the contrary, turns faith against fidelity in favor of belief. The individual who receives or captures messianity becomes a noema or a clone of the old individual as believer in the world, a clone under-Christ or in-Christ who strips man of the old belief.

In relation to belief, faith proceeds through a reprise of superposition, and this reprise is not a reflection, it is a new immanent ascending, a vector that subtracts itself from theological representation. Messianity, and thus faith, is positive, and possesses its own actuality as immanent praxis in relation to belief. As for belief, it is ecstatically collapsed into the world. The death of Christ perhaps signifies the sacrifice of all worldly knowledge; to "believe in him" is to be faithful in-immanence to his abased transcendence. Faith, under the effect of superposed messianity, is always a subtraction of macroscopic belief from itself. It has an object and also manifests religion as faith in a trompe l'oeil, wherein the objective appearance of faith is grasped spontaneously and doubled onto itself. To think faith rigorously, first with rational and algebraic instruments, and then to transform them through a generic or radical becoming—these are the two stages necessary for the immanent opening of messianity to give onto the noema of faith that replaces the old individual ego. The deindividualized ego, traversed by messianity, is the faithful-existing-subject.

FAITH, THE INDETERMINATE KNOWLEDGE OF MESSIANITY

Is it a question of a quantum-theoretical secularization of faith, after so many other secularizations? But many secularizations are merely critical and ultimately materialist reductions that consist in a refusal, without the least recognition of theo-christo-logy and its own type of positivity. The quantum-theoretical and generic reduction of faith must be capable of generating at least ideally the orthodox or Christian forms of faith and its institutionalization in the Church. This is why our enterprise ends not with the experience of a pure and simple negation of Christianity, that is to say, of belief, but with an objective uncertainty of faith that we oppose to belief, our material, an experience oscillating between the demand for fidelity and the indeterminacy of faith at least as cognizance of messianity or of grace.

Messianity, even if it ultimately conditions it, cannot be its own cognizance qua divine "plan of salvation" (with its conflicts between freedom and divine prescience), as the theo-christo-logical circle necessarily must conclude. The distinction is, more generally, that which is implied by an experimental process of knowledge in which the real object to be known and the object of knowledge produced at the end of the process cannot be philosophically confused with each other. The phenomenal or real content of faith can be none other than Christ as messianity. Faith is the reprise of messianity, and it also divides itself in two directions—either it is diverted from itself as belief and closes itself up in objects, acts, or speech that it intends in the world that is, in any case, the noematic content of faith, or it only intends them indirectly, as does the generic subject that it is in-the-last-instance.

If the knowledge of messianity remains indeterminate or is not firmly assured of itself or of being "faithful," it hesitates objectively between faith and belief, which are the ultimate opposed becomings of faith and cognizance. The Superposition realized by the Ascension and the Resurrection, which are a reprise of messianity, permits a dissociation of belief and faith, even if belief is a faith diverted from itself, perverted by theological transcendence. By which we understand that the disaster that is the Church, attached to the conservation of faith as the belief that it holds together with an armature of dogmas. Faith is given us by messianity, but this gift is the greatest of risks—that of the reception that does not understand, or that reduces itself to its worldly conditions of existence. For the faithful or the "newly converted," the danger is that of not understanding one's own faith or absence of faith, about which one can interrogate oneself, and the conditions under which it is given or indeed refused, of interpreting it as completely determined and assured, whether present or absent: ultimately, the danger of not knowing that, as faithful, one is implicated in messianity, rather than being its believing observer or its theologian. The Church is the subject of antigrace, the grasping reception of a gift that it persists in wanting to deserve. Such is the self-justification of the works that have diverted faith from its immanent work. Faith as fidelity admits this ultimate uncertainty, and as far as possible avoids reifying it in dogmas of belief or of atheism. Owing to its status and the conditions of its production, faith as cognizance oscillates between aleatory or theoretical probability and the worldly contingency of miracles. If messianity is vectoriellity, faith is the probability of the presence of the messianic message realizing itself and being received by individual subjects. From the point of view of theology, and still, in part, that of faith, messianity remains vague, detained, like the undecidable essence of Christ, between the pure immanent phenomenon of faith and the corpuscular object of belief—it remains ambiguous. In particular, christology claims to discern the predicates of Christ or to know messianity, but in doing so obtains only that which corresponds to the instrument of reception. One cannot define or interpret messianity in Greek or Jewish terms, before having reprised or lived it; to treat it as received or present in the world is to extract it or withdraw its entire essence and replace it with the transcendent religious form that is belief. One cannot at the same time identify the sources of Christianity and think messianity or Christ correctly according to the practical ignorance of its essence and its becoming. The Christian-Religious type, the man of belief, claims, with a certainty peppered with a grain of salt of doubt, that he has or does not have faith. But to "have it" *in all uncertainty*, one of the faithful must have "rechristened" his life, operated a reprise of immanence that makes him generic human or Christ. There are qualitatively different uncertainties: the uncertainty that comes from the lived essence of faith, of messianity, and that hesitates in uncertainty rather than in ignorance of itself; and that of the theological-historical knowledge of Christ, of the belief that *believes it must be in a position to* solve all problems. Believer? No. Faithful? Yes, but fidelity does not assure us definitively that our faith will never again be a belief.

THE UNILATERAL DUALITIES OF VECTORIELLITY

The christic stance defined as vectoriell and generic testifies to a curious, complex, syntactically nonphilosophical duality, one that we might call "unilateral."

I. *The components, idempotence and the lived, neither foundation nor subject, of vectoriellity.* On the one hand an objectivity of a type unknown in philosophy, a strictly algebraic idempotence; and on the other hand, fused with it, a lived that is inseparable from it, their hylomorphic combination constituting the substance of those vectors that might be called either messianity or faith, depending on the level of analysis—in any case, a generic property of humans not as religious beings but as beings who know their faith as nonreligious messianity. Thus a scientific rule and nothing else; not a metaphysical foundation or a first cause, but a scientific principle of "transfinite" continuity, the idempotence of algebraic origin. And a lived neutralized by its fusion with the scientific procedure—not a subject, but a lived that *could* be said according to the domain of objects, either faithful or of faith, if not of belief, if it is a matter of a science of religions, or an amorous lived if it is a matter of a science of eros, or a creative or artistic vector if it is a science of art and aesthetics, or an vector of action or an act if it is a matter of a science of power. There is, for example, a quantum of faith-without-belief or a quantum of acting-without-action, a neutralized belief or act. In all of these cases, this lived has nothing egological about it. It is a part of a protocol that constitutes the subject-science and paves the way for the procedure that generically neutralizes it. In short, there is a lived that is neither that of a whole, such as life, nor that of a singularity, such as a particular living being, but that adheres to a scientific procedure. Thus, the composition of a vector requires a law of immanent objective consistency *plus* a quality drawn from the domain of the objects treated, but generically neutralized.

2. *The aspects of vectoriellity, messianity and faith.* This fusion of two indissociable components gives rise, depending on the degree of analysis, to a duality of aspects in the so-called unilateral form. On the one hand, the two components of every vector form a flux that is immanent through and through—this is messianity properly so called. On the other hand, it opens up and has complex repercussions (unilaterality) upon the lived from which is drawn the subjective side or the faith side of messianity, and which thus has an origin or provenance other than idempotence as law of messianity. (A) From the point of view of messianity, the "Christian" lived of belief changes in nature and finally becomes a faith without-belief. (B) From the point of view of faith, which can be isolated under condition of the world, a duality retroactively emerges, that of faith and of messianity, which now can in no way be confused with each other. Messianity thus risks being understood as a projection or an image of faith. The constants of generic disciplines always have two unilaterally discernable destinations that are deployed in one direction, toward the world, or contract in the other, toward messianity. This other duality, unlike that of the components, is not entirely traced from the exteriority of the world. Under a now determinant messianity it becomes "ambiguous," with two possible interpretations. On the basis of this ambivalent situation of faith, which is an effect of nontheology, but not at all its essence or its "last instance," we can understand how easy it is to pass from messianity to the world, through the knot of faith, how the grace of messianity can flip into the hell of theology, never to return, or end up making messianity the genesis of this hell. This is how generic man comes once again to be gripped by the original sin to which he consents via a confusion that can itself also be understood in-the-last-instance or generically. There is ultimately a double duality (messianity/faith, messianic faith/faith fallen into the world or belief), given the fact that the unilateral duality extends as far as the world as the domain of theological, religious, and philosophizable objects, which leads its sufficient existence independently, while being affected by immanence. In other words, Christ as subject-science is not commutable

with the Christianity it may generate; this, moreover, is what allows us to treat Christianity and its theology as mere materials for this science.

3. *The generated duality, faith and belief.* The duality of messianity and faith is extended into exteriority by that of faith and belief, which is faith cut off from its origins, a conclusion without premises, encysted in itself and positing itself as sufficient. Belief is the great confusion entertained by churches and sects; it is the outcome of the confusion between messianity as origin of faith and the world. If we compare it to ontological difference, it is parallel to the confusion of the meaning of Being with that of beings, admitting the distinction, of course, between Being and the One-in-One of messianity, between philosophical and generic contexts. The genealogy of belief on the basis of messianic immanence and the "noematic" faith that inhabits it renders intelligible the phenomenon of belief that we have had to admit a priori, but without any vicious circle, as "theological symptom," in order to be able to treat it as mere material. This "reification" or "substantialization" of faith into belief is the destruction of the constitution of Christ as superposition of states in every man. Rather fidelity of the last-instance than faith, rather the faith of the faithful existent-Stranger than religious belief. If there is a guiding slogan to this work, it is indeed the former that makes the "Christ" matrix and its dualities function. The invention of Christ is nothing to do with personal individuality; it has an irreducible double aspect, which can be opposed to every unitary and/or trinitary theology precisely insofar as it requires it in order to constitute itself.

REPETITION OF THE MESSIANIC WAVE FUNCTION

The Christ-science is constituted of many strata, concatenated like phases. (1) Christ as generic subject is first constituted by a formal principle of algebraic origin that is used in quantum science—the property of idempotence or the form of a vectoriell immanence, a flux of immanence that is the form of all messianity and generic grace. (2) This flux of messianic immanence is not an empty logical form. It has an intrinsic idempotent content in the form of a materiel or generic lived. (3) Their ensemble, that is to say, the superposition of the scientific element and the lived, constitutes the vector of transfinite immanence, the messianity that is for itself, without ego, like a nonindividual subjectivity. But it is necessarily accompanied on its worldly pole by a messiah or a faithful, of individual or worldly origin as is Jesus himself qua man, but transformed into messiah-existent-Stranger, stripped of all belief and all individual transcendent ego—this is a unifrontal or unifacial faith, a faith without opposed object and without individual subject. Thus the immanent messianic vector possesses not only a form (messianity), and a content (the materiel and messianic lived), but also an intentional content (the faith that is in its origin alone doubly transcendent). (4) This doublet-transcendence is the general form of the religious sphere, a corpuscular form of the world (beliefs, rites, meaning, symbols). It is itself reduced or split, it has a double status, at once extrinsic or in-itself and hence apparent, and also immanent precisely as messianic or faithful ego. It is therefore a transformed transcendence, simplified or unilateralized by the flux of immanence that traverses it and in which it is rooted. But it has a "double" that is the grasping of it as a sufficient or in-itself thing—that is to say, belief.

It is important to distinguish, in the form of their unilateral duality, messianity, which is not a historical or theological abstraction but a generic or real essence, from the messiahs that it configures and guides at a distance and indirectly in their fidelity. Almost all theo-christo-logy is founded on the unitary and amphibolous conception of the Messiah, on the identification rather than the superposition of messianity and the messiah in the religious notion of "messianism." Messianism betrays its neutrality and its philosophical anonymity; it is a mere historical coalescence of Judaic notions reappropriated by Christianity, a generality that can withstand every possible indignity of textual deconstruction (messianism-without-messiah). Its quantum deconstruction is not founded in a metaphysical dualism but in a weakened, unilateral duality, which agrees with another principle, the immanence of messianity proper to the generic or to its reprise. The radical critique of unitary theology and of its plan of salvation consists in "extracting" methodically, by way of the so-called imaginary Christ-factor, the original christic contribution (with which it cannot be wholly conflated) from the milieu of double—that is to say, ultimately, Greek and Jewish—transcendence. Through this, we gain the possibility of an explanation of christo-theological difference by dualyzing this macroscopic unity, introducing the Christ-vector-form that is not macroscopic reflection but quantum superposition. Faith must also be quantified or "measured," torn away from the bad indeterminacy of the theological imaginary. In other words, we are dealing with a passage from the psycho-metaphysical imaginary to a more rigorous imaginary represented by the negative quarter-turn. The latter is the Christ-index necessary to messianity as flux in a nonworldly or nontheological space. For there to be superposition or interference but not identification, and thus for there to be vectoriellity or messianity rather than the imitation of an external, transcendent figure, we need the Christ-factor that forces the real into messianic wave-immanence.

Five

ALGEBRA OF THE MESSIANIC WAVE

THE IDEMPOTENCE OF THE MESSIANIC LIVED

Idempotence (A + A = A) is the "protologic" of superposition and prepares its own superposition either with the microphysical world of waves or with the lived human world. As elementary form of superposition, it is capable of soaking up the lived, of forcing it, finally succeeding in this exploit of fusion—in short, of being not just immanent but immanental, or of being valid *for* the real of the subject. But how can a mere algebraic property suture itself (and more than this) to transcendence? Various solutions might be envisaged.

1. A given algebraic property in a science has no chance of encountering a lived subject as such; no suture is possible between them unless we presuppose the problem resolved, unless we presuppose in one way or another a suture, for example, by treating them as parallel spheres, whether specular or linked via torsion, one ontological, the other evental and already adumbrated in the first, thus returning to the fold of philosophy in the form of a generic materialism.

2. Given as manipulable and "calculable" in and by a machine, the subjectivation of the property will depend upon the complex nature of the machine. One might perhaps speak of a "subject," or even of a "lived"—as one would speak of it not in terms of consciousness, but rather in terms

of a lived of man, understood philosophically and religiously *since it is the material or the object that determines the relevance of our project.* We are not bound by consciousness but, let's say, by the practical subject qua act, a lived in a wider sense than the liveds of consciousness. There are amorous, artistic, practical liveds, liveds of affectivity, of religious belief, and so on. It is the lived in its generality that interests us.

3. Assumed as given in the world itself, algebra and the lived have some chance of encountering each other as transcendental fusion in and through a subject that is a captive of the world. We eliminate this purely philosophical solution, for it falls simply under our object or material.

4. Having eliminated these solutions, which are irrelevant, for various reasons—but in general because they fall under the transcendental interiority of a philosophical project—are we left with no solution, obliged to recognize a miraculous event in Christ?

Algebraic idempotence must be interpreted or translated philosophically and/or theologically in order to be utilizable in the matrix. A phenomenologically "discreet" description, which certainly makes use of concepts that clearly of philosophical extraction but that play a nonphilosophical role or fulfill a nonphilosophical function, is possible. The science we seek here is that of philosophy and theology as object; but we know that it must include an aspect of the latter in its apparatus. The science of philosophy is more complex than its strictly scientific means—by definition, we suspect—and the problem lies precisely in the possibility of this complexity. A science that constitutes itself is more complex than the local means it draws from other already-existing sciences. And what is more, no science can be reduced to its axiomatic, which is always precipitate, never sufficient to it, for it remains open to an outside—here, open to new mutations or new statements of philosophy. Abandoning the ideal of the axiomatic-all, we are obliged to admit that this science can only present its two major principles of quantum-theoretical extraction by interpreting them first in terms of the subject or the lived, which is thus supposed already to be included, albeit nonthematically, within the principles themselves. Superposition and noncommutativity cannot be mere "brute" algebraic properties, but are raised to the state of scientific *principles* and must be understood as such. The idea of a science of religions borne by Christ is indeed a miracle of the highest and most straightforward order, like any scientific discovery that exceeds its axiomatic. But we must push as far as possible the analysis of the conditions of this immanental, and not merely transcendental, miracle. This is what we mean by saying that the axioms of this science are "oraxioms." Idempotence can describe itself indirectly or in-the-last-instance with philosophy as a silent or semispeaking subject, underpracticing the concept. It is then described as strong analytic and as weak synthetic, or as mediate-without-mediation, as generic mi-lieu, midsite, which is not a median between the religious components of a science, but precisely the unilateral duality characteristic of the generic. As strong analytic, it fuses—such is its ("weak") "force"—with the subject as external given, but on condition of reducing it analytically to the lived *alone*, or reproducing it analytically solely as lived. As weak synthetic, it fuses with the subject but while adding to itself the lived alone, without adding to itself the philosophical ego, which falls from its great height into immanence. Fusion via idempotence requires both aspects, analytic and synthetic, under the form of their unilateral complementarity.

THE PREEMPTING OF THE LIVED AND THE MESSIANIC WAVE

The operation of the matrix is entirely immanent, even if it puts into play on one side of itself the most exacerbated theological transcendence, which it makes fall into-immanence. For there is never theology, even a theology in-Christ rather than in-Philosophy, without a transcendence now unleashed into belief, now abased into faith. Generally speaking, the matrix supposes a vectoriell reduction (to $\sqrt{-1}$) of the onto-theological and doubly transcendent One. But its effect is divided according to a duality or unilateral complementarity of aspects that flow from the transformation of the metaphysical One, always marked by a duality insofar as it becomes involved with philosophy. On the one hand, the One is now superposed with itself—it is what we call the One-in-One or the radical immanence of messianity. On the other hand its second aspect manifests itself as transcendence, but this time fallen into-immanence and simplified, contributed by messianity—this is what we call faith, or the faithful-existing-Messiah. It will be remarked that the matrixial conjugation and underdetermination of the variables given does not make of them a simple semantic table of various combinations, but a machine productive of the Christ-science with its two aspects, messianity and fidelity.

Its principle is quantum but generic in essence, essentially the superposition of idempotence, included or programmed in the latter, and of the lived preempted over phenomena of belief and transformed into a messianic lived. This conception is essentially concentrated in the phenomenal immanence rendered possible by superposition. We call "wavelike," in general, waves *of every nature* that possess the property of being an algebraic structure of so-called idempotence. The wavelike is not said solely of material phenomena that pass for "hard"—water and its waves, the earth and its seismic movements, materials traversed by internal movements—but is just as valid for the sphere of decision, and that of affectivity and art, obviously. Just as even the hardest glass imparts a quivering to things seen through it, a fortiori there are quiverings of the soul, wild oscillations of the heart, a periodic swaying of beliefs and of opinions, great theoretical migrations and comings-and-goings, not just paradigmatically isolated summits. There is nothing to stop us from conceiving of musical waves of belief, and hence of a generic messianity as wave function or state vector of faith. Although to speak of a physics of messianity, of spirituality, and of faith does not scandalize us, as we shall see in the conception of the Cross as crucial experiment, too many misunderstandings are possible here; and yet theology may educate itself with the physicist school of mysticism, even the most Neoplatonic among them. Materiel, quantum, and generic physics has nothing to do with a "physicalism." All of these phenomena, if they are to be understood and rendered intelligible as wavelike, must obey the property of certain operations of being idempotent, rather than being reduced to an inconsistent psychology. Idempotence is more or less distinct, but is not identified as such by perception, nor even necessarily by the equations of "wave functions." But no christic science would be possible without this minimum of algebra. So statements such as the following should come as no surprise: that the messianity of Christ is of two "states," the Greek state known as Logos and the Jewish state known as Torah, and that our task as quasi-physicists of faith is to establish the state vector of the christic state of faith, the wave function of a human or generic messianity.

THE SAME OF IDEMPOTENCE

An operation is therefore idempotent if the same term or the same component can be added or taken away from it without its truth-value changing. It remains the same whether there is addition or subtraction in the result. Idempotence gives immanence immediately in its linear and "wavelike" form, not in the form of a point or a circle, as philosophers suppose. It programs the identity of a term with itself as result or resultant, whatever may be the operation that serves as its mediation and that is therefore neutralized or suspended without being, for all that, forgotten or negated. The intermediary operation, which we might have taken for the transcendence of an arithmetical addition, is suspended in its effect. Idempotence simulates the identification that makes one term fall back upon another entirely, and makes them see each other in the mirror or according to the Moebius band. Whether total or even partial, identification remains an operation that overflies immanence, neutralizes nothing, or destroys everything. On the other hand, out of two terms—here, two Laws (Logos and Torah)—superposition makes one, one sole Law as interference, but on condition that it is immanence, of a lived nature, and wavelike, and that the mediations between them are neutralized, not suppressed or denied. Idempotence is a linear immanence, a Same that has as its condition the suspense of the operation of addition that would be understood as arithmetical and would posit terms in a space of exteriority or transcendence. Two entities in a theological space are not superposable, whereas an idempotent addition produces a "same" as flux and phases of flux. Idempotence signifies that the messianic lived is constant whatever term may be added to it; for this term, by being added or adjoined to it, falls precisely under this constancy to which it contributes a "supplement," but which it leaves constant—it is therefore the law of a sterile excess that does not destroy the Same. Messianity must be understood as the unicity of superposition or addition of self to self without its nature changing. It is the vectoriellity of the Same that remains the Same as itself. It is not an empty algebraic form, but is capable of fusing, given its superpositional nature, with the lived property of subjects. As force of fusion, it finds its immediate matter in a form of the lived that belongs to religious beings, belief, which it preempts, and from which it samples the generic form of faith. Torn from belief, faith is of the lived, which is neither object nor subject, since the two poles are mixed in every belief. Messianity is "materiel" right up to the most extreme spirituality, lived and not empirical, as a constant of faith that remains the "same."

UNIFACIALITY OR SATURATION OF THE REAL?

The second principle is that of the noncommutativity that brings about the necessity of a certain order. It assures the being-oriented or the sense of the "last instance" of messianity, and this by means of its being-foreclosed to the world, messianity not being commutable or exchangeable with any phenomenon of the philosophically intelligible world. The world is indeed first, but messianity is prior-to-first, and underdetermines priority. It transforms transcendence but is itself without ecstatic transcendence.

Idempotence is perceived under two aspects, like a relation that remains analytic despite all the complements that are contributed to it and that merely make it strong analytic and open, and like a weak synthetic or completed relation because the complements contributed do not modify it and complete it without closing it. This is a phenomenon that has no site in its terms, and that is a phenomenon neither of contradictories nor of opposites, as if it only frequented the margins of established and corpuscular things; it is not even "marginal" like the border of an instance or the exteriority of an interiority, like a nostalgic wandering. It frequents not the borders that face backward into the interiority of transcendence, but the very structure of the border or face insofar as there is no border-to-border or face-to-face. The border is a One-border, a uni-border, the face is a One-face, a uni-face. The idempotence of messianity is the pure bursting out of the face or the border, their throwing insofar as they have lost all reference of departure and arrival, the pure thrown that has no discobolus but that is facing. It is the uni-face that is the face-(of)-the-One, and the One is certainly not a subject that throws, but the movement without trajectory whose sole trace is not even left by it, but left as unique face.

Idempotence defeats the phenomenological distance inseparable from a subject opening or thrown "before itself." The adjoined instance is thus not adjoined at the end of an annihilating distance that would, in any case, be destructive of the idempotent "packet" of liveds. Nor does it end up saturating this distance with too much sense, or transcendent alterity, with an Other either gaining a foothold directly in the lived, or manifesting itself in the call. We recognize the philosophico-religious problematic insofar as it functions on the invariant of phenomenological difference, with outbreaks of opposites, of more and of less, of saturation and of desaturation, whereas the science we require as a model functions on algebraic idempotence (and the imaginary number for vectoriellity) as engendering superposition and noncommutativity. Idempotence is a saturation but a sterile one, we might say—it is a radical immanence and not a transcendence *necessarily doubled if not reduced or simplified by its falling into-immanence*—always the key problem for a complete analysis of philosophy as a transcendence in doublet. There is always an excess, but the transcendence of the lived phenomenon is transformed, the simple being here an excess via superposition rather than an excess via saturation. This is the whole difference between a religious phenomenology of the limits of transcendence and a quantum theory within the limits of the immanence that depotentializes theology. The ego is superposed generically rather than interpellated or placed once more in excess over itself by an excess of transcendence over itself. The problem of intersubjectivity is no longer posed for it; without being many or solitary, it is underdetermined in its relation to the world that has become its sole interlocutor. The couple Ego/Other has been traversed by the Same and redistributed according to a unilateral complementarity. The subject cannot be interlocuted or interpellated because there is no dominant interlocutor, but only a scientific Other as generic Same, an immanent body of Christ if you like, which charges itself with leading subjects to a revelation—that is, to faith.

IDEMPOTENCE AND THE TWO QUANTUM PRINCIPLES

Idempotence allows us to rediscover the two scientific principles that, intimately bound together, submit belief, transform it into faith, and produce the human intellection of Christianity.

Idempotence is that common root of the two principles required of faith or messianity, in order for it to become a constant, a certain type of determination governing the relations that it puts to work. It is an algebraic property or a proto-principle. Its first effect makes it function concretely in quantum physics as a "principle of superposition." An idempotent operation produces the same result whether it is applied once or reapplied many times, for example, in the form of the flux or forces of the lived. It thus produces the unicity of the intermediary (what we shall call the "mediate-without-mediation"). It is half-analytic, half-synthetic, or more exactly, it is the analytic no longer as isolated or corpuscular philosophical operation, an analytic that "tends" to the synthetic or aims at it. Complementarily, it is the synthetic that fails to be as it is qua isolated or corpuscular operation in philosophical logic.

The second effect of idempotence is the principle that subtracts this determination from its always active or threatening philosophical reprisal. The algebraic property of noncommutativity is raised to the status of a principle in quantum theory (for which the inverse products of two variables are not equal), and here to the principle of the noncommutability of messianic faith with religious belief.

Analytic and synthetic, thus delivered from their philosophical combination, reduced to their operatory algebraic kernel, together form what we shall call a *unilateral* (and not "dialectical," as in positive physics à la Bohr) *complementarity*. The immanental principle par excellence of idempotence breaks from the start with classical logic and the philosophy that it supports in "relations" between science and philosophy—in this case, that of objectivity and of the lived of belief, that is to say, in the constitution in-the-last-instance of faith. In a perhaps too condensed formula, idempotence is the messianic criteria of faith and that which distinguishes it from belief. It is still necessary to fix more precisely the type of real or of lived that makes messianity and its vectoriellity out of idempotence, and tears it away from the world or from "reality." But in any case, built on these two principles as upon the pillars that make for generic humanity, messianic faith is the condition of the ruin of theological sufficiency, and the condition of the intelligibility of religions. As difficult as it is to recognize the affinity of the Christian invention, for example, with the Marxian foundation of a science of history (Michel Henry), its affinity with a physical science like quantum theory is yet more surprising. Still, we have distinguished in the christic operation these two principles from quantum mechanics that give it its generic universality. They are both decisive for the definition of messianity as vectoriell.

THE GENERIC CONCEPT OF THE UNIFICATION OF THEORIES

Idempotence contributes toward defining the generic matrix, it plays a part in the superposition of science and philosophy; but in order to be effectuated for the two disciplines, it must itself also be incarnated in philosophy or in the lived of a subject. Generic means a nontotalizing validity for the two disciplines, or a universality that does not double back on itself, which disappears even as reflected All, and is therefore an indiscernible universality. The key to the problem lies in the idempotence of science and philosophy, which are the same generically or in-the-last-instance. We finally have the rigorous concept or the equation Unified Theory = the Same as idempotent superposition. We obviously oppose this matrix to that of Parmenides, the matrix that structures philosophy. Any science or philosophy whatsoever can add itself to itself, but only on condition that this takes place under the principle of superposition. But the latter is usually replaced by a transcendent principle of identification, and this type of unified theory thus remains on the order of the symptom. Only quantum science in the order of physics, and the christic science of religion in the order of human sciences or of theology, satisfy the immanental rather than transcendent(al) form of unified theory.

The most general principles of science, those that pass over frontiers between disciplines and even between theories within a given discipline, are of a mathematical and in particular an algebraic order. In philosophy, they are only transcendental principles, but when they become immanental principles, quantum physics is endowed with an intelligibility amplitude higher than that of classical physics. What is more, they are transferable, qua principles of a nonpositive quantum theory, to nonphysical properties, but ones that possess a certain materiality, such as religious phenomena or the liveds of belief. If a science must be possible in-Christ and valid for all religions, it will be realizable with these operations which, taken together, form a rationality that is not logical but, as we say, "quantial." It is important that these are principles of great amplitude, and therefore algebraic, and that in order to become principles they must be detached from physical properties and must be transferable into another context of objects, into the matter of faith and of religions, without for all that being universal in the philosophical sense, universal via hierarchy and domination. Marx, in his own way, had an inkling of their use for the science of history, with his theory of determination-in-the-last-instance. This is the example of the fusion of (revolutionary) theory and the (proletarian) masses—or, for us, that of theology and the faithful of Christ.

Idempotence is the scientific matrix of the generic. It is valid for the superposition of science and philosophy, but remains to be effectuated or realized in the two disciplines. Science and philosophy are the same generically but only in-the-last-instance—this is the rigorous concept of unified Theory. Kant, for example, with "the same" of the experience of the object and of its representation, remains in the state of a symptom, as indeed is indicated well enough by the way he internalizes Newton into classical metaphysics.

NO RETURN OF CHRIST, BUT A CHRISTO-FICTION

The first effects of the two principles' investment in our problem are visible as the transformation of theology and the production of a christo-fiction that replaces the "return of Christ" as object of belief.

1. The superposition of Christ (of his sayings) and of Greco-Judaic theology, the addition of christic sayings to theology, must yield the kernel of the christic discourse, but in its theological version. Whereas inversely, the theological discourse is transformed, folded, or ordered according to this principle or operation.

2. Why is it not a simple reciprocal, specular imitation, a simultaneous transformation; why does Christ become law and in this operation remain the same without letting himself be affected by theological discourse; why is there a transformation of Judeo-Greek speech alone, in order to become adequate to the person of Christ? Another principle accompanies that of superposition, giving it its sense and its limits, making explicit its consequences—it is the principle of the noncommutativity of Christ and Judeo-Greek philosophy or theology. In other words, either all theology is of the order of the philosophico-religious imaginary, an imaginary formation that serves as our material, or else, if we elaborate a new formation as a function of Christ and of this imaginary as material, it will not belong to that imaginary but it will thus remain without any traction on the foreclosed Christ; it will be exactly what we might call a

christo-fiction. Theology, and Christianity in general, are thus involved in a transformation more profound than any reformation. It is what we call the unilateralization of Christianity, which ensures, on the other hand, the defense of Christ against Christianity. Noncommutativity allows us to understand the foreclosure of messianity in the course of the world and of history.

The Christ-event is the emergence of a new science, the science of the world as object of belief, a world of which Being, the One, and the Idea are the modalities. The world is the gnostic object par excellence, with or without globalization. The confusion of faith and belief, of messianity and of the world, is the great disaster that affects the faith revealed by Christ, and of which Christian confessions are the agents and the consumers. Of course, this destructive confusion is the real content of "original sin," which is its mythical projection. Just as lived and subjective faith does not belong to a subject but first of all to a subject-science, faith does not refer back to an object, even an interiorized and idealized one; it is of an immanent nature, which is neither an all nor a singularity, but a superposition. And superposition is messianity as fidelity to/of vectoriell immanence.

FROM INDIVIDUAL RESISTANCE TO GENERIC DEFENSE

The messianic wave contains no atomist or corpuscular intuition: the particle of the message or the kerygma is not even "partial" but instead "quartial," as we have explained elsewhere. It is all of classic theological representation that is inadequate; fallen back on messianity, it destroys its principle of idempotence. Radical messianity is a wavelike flux that excludes the individual and the all, the symptoms of divine transcendence. Idempotence is that matrix of a linearity that is algebraic rather than geometrical. It suspends the reciprocal mediation of God and man and makes no circle. One can add to it the same term or an Other since it brackets out the operation without denying it. In consequence, the sacrifice of God is programmed in the form of his neutralization, his suspended presence. Idempotence is a term's power to remain the same across, through (not *in spite of*), the crossing of the operations of which it is the object.

But in relation to history, the under-going Christ introduces a great disruption. Messianity is no longer that which remains or resists, and whose stubbornness defines the Same as residue of History; it is the Same that resists or remains because it is prior-to-first or the last-instance. It is Christ that is our fortress, our reduced interior, and unleashes the destructive fury of the Adversary. The variations serve to make appear that which resists—an operation of glorious and stubborn transcendence. Whereas here the Same does not have to constitute itself as that which resists; it is content to move in itself and to oscillate at the whim of variations, operations, and differences, not in spite of them. The Messiah has the sovereignty of the weak, a welcoming or unmoved Indifference. The nonact of the Idempotent does not enter in a body-to-body or a face-to-face, does not have to react reflexively in order to resist. This power of being and of remaining the Same without having to kill the Adversary is messianity. It is not just the power to resist variations and avatars, to include or exclude them, or at best to make a synthesis of them, in a great theologico-worldly system. It is what distinguishes, from our point of view, the defense of humans from the resistance that can only be a combat whose horizon is death. Neither affirmative nor negative, defense includes a moment of transformation, rather than the putting to death of the Adversary; it does not, for example, have to constitute itself through the diasporic or unitary triumphant resistance of contingencies, including them by negation in an affirmative diaspora (just imagine) like an Eternal Return of the Jew, or in a system designed to pick them out. From the Same of philosophy, which constitutes itself as surpotence and omnipotence, to the Same of messianity constituted not as power of return or of repetition but as the impotence of idempotence. The Same of messianity is not placed between two parentheses through an act of constitution; it is constituted by one sole parenthesis and thus is open to the transfinite.

MESSIANITY AND THE FAITHFUL, THEIR UNILATERAL COMPLEMENTARITY

We have distinguished between various significations of the symbol "Christ," as message-system and as factor-(of)-fiction or "imaginary" factor. A third modality that conjoins these two is the Christ-operator or agent. In every science there is at least one operator-observer of the research procedures that are the true "subject." In the positive sciences this functional subject is external and is an "agent" of the laboratory, the physicist as operator or "preparer" of corpuscles put into a quantum superposed state. But in any case it is not the subject-cause of science. In disciplines like psychoanalysis and Marxism, and as tends to be partly the case in philosophy and the truly "human" sciences, a fortiori, a functional operator is required, but is not the cause of science here either: it is an individual nonfree "lived," placed under generic condition. The lived is at once knotted together with the scientific project or principles, and at the same time and in complementary manner also an effect of their power of configuration. A subject that remained in itself or macroscopic, as a subject of the operation of measurement or of constitution, would instead destroy the physical superposition, just as a priest would destroy the reality of messianity by transforming it back into belief in the form of the world. What matters, therefore, is that some lived, some operator or manipulator of variables, must be superposed with the two principles and included in a unique state vector. What to call this subjectivity or this individual faith that is underdetermined generically if not "messiah" or subject-existent-Stranger? A science, even a christic or "non-Christian" one, must eliminate the philosophical subject-foundation, must be a true science that recognizes its contingency, a procedure whose objectivity cannot justify itself through a program of autofoundation. And at the same time as a science of human stances (religious stances, for example), it must include a subjective complement to the procedures. The Messianity of-last-instance and the messiah of individual origin, but stripped of its individual predicates, forming a unilateral complementarity.

Certain sciences, perhaps all sciences, merit subjects that are included in them rather than being solely their object—sciences that introduce constants into the intimacy of thought, a constant of salvation, an indivisible messianic ultimatum. The "last things" are not fabulous, cataclysmic cosmic events. Only generic subjects (rather than those who return) are "last" in the eschatological sense, that is to say, "prior-to-first." The name of Christ does not introduce a fundamental discontinuity "into" the present of history, but something of the "futural" prior-to-present into this present of history itself. Far from "breaking history in two" (Nietzsche) according to the old philosophical schema, it brings to light a prior-to-priority that allows history and religions their priority but places them under underdetermining condition in-the-last-instance.

Rather than reducing Christ to the rank of any faithful whatsoever—that is to say, a "Christian" or "believing" one—we shall admit that Christ is not "Christian" like Jesus, in that belated sense, still less a "believer," but that, on the one hand, any one of the faithful whatsoever must be put on an equal standing with Christ, equal among equals, freeing himself from the Christianity-world, and must be a modality of generic science or of the faithful of-the-last-instance, not subjected to messianity but generically subjectivated by it. All of the faithful break as far as possible from the yoke of Christianity instituted as religion, the yoke of the vicious circle.

A subject of the classical form—ego, consciousness, or singular ipseity—there can be nothing of this in the human sciences if they are really to be sciences, and in particular in the science we seek here. Faith excludes them but does not deny them; it transforms them. But on the one hand we have subjectivity in the interferent or wavelike state, and thus on the other hand we also have subjectivity in the deindividualized form of messiahs expected with certainty but with indiscernible trajectories and unforeseeable effects. They act without being localized in predetermined theological spaces and times, entangled messiahs proceeding not through encroachment or through a play of frontiers, but through effects of grace at a distance that go through the obstacles of the world. There is no subject of science, but only a subject-science that, invested in the givens of worldly Christianity, produces unlocalizable multiplicities of faithful acting indirectly, not at the far end of an onto-theological distance. Such a distance reproduces philosophy and in particular the Logos. The mode of action that is opposed to this confusion is probably essentially vectoriell. On the basis of philosophical subjects as believers, messianity configures messiahs—but without a relation of mechanical or dialectical causation. Messiahs are vectors whose essence is onto-vectoriell rather than geometrically vectorial.

Six

CHRISTIC SCIENCE AND ITS OCCASIONS

THE GNOSTIC FUSION OF QUANTUM SCIENCE AND THE SUBJECT: FUSION AND SUTURE

Why quantum science? And why fusion rather than suture? The Greek philosophy that served as a milieu for the reception for the christic message, the Logos, is not simply a question of various objects, themes, ideas, or concepts, but is a representational form of thought that applies to almost all of its concepts, including those of theology. The deconstructive scrutiny of representation undertaken by philosophy itself is a necessary but unfinished task, which has still not been posed in the most effective terms. To think an event such as the messianic subject, other means and forces of production are needed. Idempotence means that the messianic lived is constant, whatever term may be adjoined to it as Other, since the term to which it is adjoined or added falls precisely under that invariant, to which it contributes only a complement. Such is the law of sterile excess, which does not destroy the Same qua messianic subject. The same goes for the diasporic punctuality of Jewishness. But it seems that quantum theory is the radical delimitation of representative thought, at least in the sciences, and its manner of organization—cosmos or diaspora—without, for all that, negating this thought. The phenomena to which it accedes are not "small" phenomena, but phenomena that obey a principle that is entirely other than that of representation. Philosophy is not equal to this other principle, however philosophical it might appear; only science is up to it. This capacity belongs to it, but it cannot be extracted from its physical application by philosophical-*type* means, or as an effect of philosophy, or imported *into* theology. It is capable of thinking religious phenomena such as flux, grace, advent, and return as taken up by the mystics and messianic phenomena in general—for example, by isolating their physical-type wavelike properties. The atomistic-conceptual style of theology cannot truly grasp these properties, yet they persist as symptoms in its heavily Greek discourse, a discourse that leaves these phenomena encysted in themselves. The notorious "metaphysical abstractions" are nothing but forms of thought that remain inadequate to their object. The quantum style possesses a generality that is in every sense ultraphysical, because it corresponds to the algebraic properties of suspense and neutralization that open up new perspectives through their characteristic type of universality.

This sort of intrinsic alliance between science and the lived is an immanent fusion, which is to say, a fusion *by means of* an immanence, and not a simple suture, in the sense of a transcendent operation merely "corrected" by immanence. In the case in point, it may be called a gnosis, a doctrine that is traditionally unwelcome amid philosophical dealings that seek to be "rationalist" and religious ones that seek to be "orthodox." It turns out that gnosis, here, is scientific, and brings a certain materiality (but not a materialism) into play. It thus turns its back on the condemnations of rationalism, positivism, and dogmatism, which, when they do not lack sufficiency, simply lack a sense of history. We do not invoke gnosis as a religious formation of meaning, although certain of the effects of this generic science have a heretical aspect—but no more so than the context of Christ, from which it is impossible to eliminate every Jewish, Greek, or, precisely, Gnostic association. The logic of our gnosis is that of idempotence, the minimal immanent flux that constitutes the subject-science, or what we shall henceforth call messianity. We interpret the immanence of faith and thought "in-Christ" on the basis of two major principles, which form a "logic" that is used in quantum science but that we could also call "gnostic," since certain properties of the algebra could also represent gnosis. It is not exactly that we seek to elevate historical gnosis to the status of logic; we simply identify a gnostic-oriented operation at work in algebraic logic: that of idempotence.

THE DESTRUCTION OF THE CIRCLE OF TRANSCENDENTAL ILLUSION

Generally speaking, philosophers begin with a combination of the means of truth and the reality of error. Some (Plato, Descartes, Fichte) insist on the error or illusion from which we must extract ourselves, while others (Spinoza, Kant, Badiou) instead insist on the truth-conditions, perhaps enveloped in illusion, that must triumph. Nonphilosophy makes use of yet another schema, which seems to combine these two types of solution without returning to the one offered by Nietzsche or Deleuze, who oscillate between the truth of illusion and Hegelian synthesis. The conditions of truth and illusion form a unilateral duality, with the former being not first, but prior-to-first or "last," and the latter being first. It is apparently dogmatic, as in Spinoza and Kant (at the beginning), and apparently critical of illusion, as in Plato, Descartes, and Fichte. "Apparently" because the conditions of these characterizations have been profoundly altered. This is a problem that is altogether worthy of christic science: how should it "begin," and what does "beginning" mean here?

Such a (non-Christian) science is not a reigning or fulfilled Christianity, instituting a science of other religions. The condition is both that Christianity be treated in terms of its religious content, as one religion among others, and that the nontheological "foundation" of this science— or, more precisely, its principles—owe everything to the operation called "Christ," and nothing, or almost nothing (only its raw material), to institutional Christianity. For Christ to be the invention of a constant that is specific to religions, or a formal trait that defines the possibility of a science, constituted Christianity cannot, in a sense, be anything more than one object among others. A "non-Christian science" of religions, on condition that we understand "non-Christian" not as a predicate but as an indivisible bloc, complementary to the subject and the law of this science. It is a matter of placing Christianity under condition of "non-Christianity," and of understanding the "non-" as the witness of science, so that non-Christian = subject-science = faith-science. The Christian or faithful subject enters into a scientific procedure and, with it, forms a constant adequate to its object. Rather than entering into science as a subject in the classical sense, the Christian enters as a Christian, but as a deindividualized or neutralized subject who can recover a generic subjectivity. Even qua indifferent material, it is necessary that Christianity be interested in its science and that it not be treated solely as an indifferent object, or as a myth.

There is the risk of a vicious circle between the Christ that is the foundation of this science and the elaboration of his meaning on the basis of constituted Christianity. Another form of the same circle: How can Christ be both the inventor of a new faith and just one of the faithful? This type of circle is well known. It is the circle of autoanalysis, whether the "Cartesian" circle or some other. In order to exit it, a study of the scientific constant is fundamental, and we must reevaluate the situation as a function of the various interlocking degrees or steps that, together, form an exit from the circle.

1. The positive scientific procedure has nothing philosophical about it, but demands the self-assurance proper to the spontaneous scientific stance, even if it be judged contingent. There is, already, this initial break from the circle by scientific positivity; already it is uncertain whether science posits itself or needs to posit itself—the very "transcendental" problem of a circle, as projected and then marveled at by the philosopher. Without being autonomous and consistent, like a philosophical subject, the initial scientific principle is inhabited by something like its own generic self-certainty, through a process of superposition, and has no need of an additional, philosophical certainty to confirm it.

2. The scientific procedure is constituted as subject-science by including and generically deindividualizing the lived of the potential Christian— one who, for example, is interested in religions without explicitly being a "believer." This doxic lived is raised to the dimension of a scientific subject, not the other way around. Science is not subjectivized under condition of philosophy. It is, rather, the lived that is desubjectivized under condition of science.

3. Introduced into the philosophical *circle* of religion and theology, and proceeding correctly, the subject-science demonstrates, through its very existence, that the macroscopic and potentially philosophizable Christian All is an appearance or an illusion through which it believed itself capable of being the sole professor and legislator of the lived.

4. There is no circle between the scientific presupposition of the science-subject, purely and simply foreign to philosophy, and Christian

discourse, or even theological science, as there is between the principle of identity or general logic and the real. This is because superposition has never been a circle, but is instead, strictly speaking, a wave or even the quarter-circle that represents an "imaginary" or "complex" number, a function fulfilled here by the symbol "Christ." The belief in a transcendental illusion present here is itself an illusion, but one that is immanental or of an altogether different degree. Philosophy is the sole positive cause of this belief in a circle, and the subject-science is only the negative cause of its manifestation. The circle is regulated by its depotentialization, by its dismissal into a particulate state and, in a sense, to a peripheral or seemingly marginal condition. The subject-science, placed under the underdeterminant or veritative condition of Christian theology, makes the philosophical or theological circle appear as such, spontaneous or specular with respect to the religions and theories that it engenders.

THE GENERIC SCIENCES AND THE HERMENEUTIC OCCASION

Human phenomena, including the phenomena of religion, demand a generic, not a positive, science. A science of philosophy, for example, requires the subject to set itself in a stance that can only be scientific and not philosophical, but that, as a stance, involves or implicates a potentially philosophizable, if not philosophizing, subject that, as lived, is condemned to pass through the objectivity of scientific principles and that, however individual it may be at the outset, is neutralized in its particular subjectivity. This does not mean that we must return to hermeneutics, that we must be Christians and believers in order to understand the faith, must be philosophers to understand philosophy, or artists to understand art. The lived experiences of faith, of art, of love, and so on are generic and are not marked by the singularity/totality of the philosophical or theological subject. The matter is somewhat more complicated: every generic lived produces a philosophizable, if not a philosophy, and this is the origin of the appearance of the circle. The lived is philosophizable according to philosophical generality, but is also underdetermined. The lived is subjectivity-without-subjectivation, without there being a fold of subjectivity and subjectivation, because it is identified with a scientific principle, assuming a certain stance and, as subject, submitting itself to an idempotent formation. Taken in its state as an ordinary, potentially Christian lived, science submits it to a uni-version, as if subjectivity were smoothed out and ejected from itself—a generic neutralization that snatches it away from both singularity and totality. The science that becomes generic thus takes up a genuine rebellion against epistemology. It is prior-to-priority, lived idempotently, even if philosophy attempts to appropriate it for its own purposes, at the risk of ending up transformed itself. It is in this sense of the lived, always philosophizable after the fact by the religious appearance, that it must have had to *have been* Christian or *have been capable of assuming* the Christian faith in order to set itself in the right stance for a science of religions.

THE TRADITIONAL CHRIST AND THE CHRIST-SUBJECT

We have distinguished various meanings of Christ on the basis of that of the science-in-Christ or the Christ-matrix, and, elsewhere, that of the historical and religious Jesus Christ.

1. A non-Christian science must eliminate the philosophical foundation, at once object and subject, in order to be a true science, cognizant of its contingency. It contains a procedure that prevents objectivity from justifying itself through itself in a program of autofoundation: the Christ-factor, the factor of christo-fiction, replaces this function.

2. As a science of human stances—religious stances, for example—it must include a lived, which the procedure borrows, and which counts as any subject's faith and/or belief, with Jesus Christ himself being one of the faithful. This unilateral complement never gives way to a philosophical All through a division that would reconstitute it. From this angle, we should treat Christ himself *also* as being just one of the faithful, we should treat the Messiah as an ordinary and nonexceptional man, but a man guided by messianity, not only as one fulfilling his apparently singular or even paradigmatic function in the invention of the principles of this science. This is the problem of the "suture" of science *according to* Christ, at once its inventor and preparer, its generic subject and its object.

3. The generic stance possesses an operative and subjective aspect, a preparation of and for experiment, an aspect of promise, and a way of rendering-vigilant that is shared with Saint John the Baptist, perhaps the prophets, and Jesus, when he demanded of his disciples that they "prepare" for the Cross, the Resurrection, and his return.

In such a science, every subject is therefore dual or complementary without being doubled in its unity. On the one hand, there is a generic lived, preempted over the philosophical subject and neutralized, carried into idempotence by its fusion with science. And on the other hand there is an original subject, the subject of the theological or epistemological occasion, transcendent or given and then reduced to the status of one of the faithful. The latter may play the role of the famous observer or experimenter included in the scientific condition. This messiah-subject or faithful is not itself *divided* into two states, microscopic and macroscopic, but the principle of idempotence draws it in and fuses in part with it, idempotence being nothing but this fusion itself, an immanental and not a transcendental operator. We shall say that the generically lived sampled from the Christian subject is "non-Christian" or faithful, which clearly does not mean anti-Christian. Along with generic messianity, it forms a unilateral complementarity, contracted onto its immanence, and a transcendence that is simple, particulate, or "fallen into-immanence."

This formalization gives a somewhat different meaning to Lacan's "subject of science," here non-Christian science. There is no point in deducing it from the Cartesian Cogito, which could not deliver it from its metaphysical placenta. But there is something of the lived in its generic status, and a complementary subject operator that helps to constitute the scientific stance. Finally, there is not, in science, a subject in the classical sense, but only a subject-science that we can invest with the givens of Christianity.

In order to clarify the relations between Christ qua subject of science and Christ qua founder of a religious tradition, let us reiterate the structures of unilaterality over Christ, in particular, no longer as generic messianity but as faithful subject and messiah.

1. *Their Christian confusion.* We first posit the distinction between the science-for-generic-subject and Christianity supposing itself in itself (Principle of Sufficient Theology), which is the condition sine qua non

for establishing an objective science of Christian subjectivity and avoiding the appearances of theology. As the "founder" of a science, here, turns out to also be—but how does this happen?—the founder of what will become a religion, these are two heterogeneous senses of foundation whose apparent unity feeds into theology. They are nevertheless to be distinguished. The "founder" of science is unilaterally implicated as objective subject within—well and truly within—its scientific procedures, which are prior-to-first. For this reason, he cannot be assimilated by any tradition he might give rise to. The founder of religion is implicated as subject-object in that which he founds, and which retroactively defines him as theological subject. It is no longer the same Christ, the same character, in these two functions, but philosophy and theology make it seem as if the two were the same, thus fueling a great confusion. That Christ should be a dual or unilateral subject may come as a surprise; it is the distinction and the complementarity of inassimilable messianity and of the "master" who is taken in by what his work has become and who is *reinterpreted* by it. Partly as incarnating the new science, forging an alliance with that of which he is a generic subject, and partly as phantasmatic object of the tradition that has been founded on it and that capitalizes on it, along with plenty of other elements. Only Christ, as science-for-subject but not as subject of science, forever escapes being the simple object of a tradition. In the Christian tradition, all distinctions between Jesus and the theologically "formatted" faithful that we are capable of being are exceedingly weak and reversible, and are insufficiently discriminating, since they have developed in the environment of continuous philosophical appearances, and therefore within a vicious circle. Subjects are always also objects for a thought that seeks to be science; but in the principles of this science gathered together under the name of Christ, the subject is also said in a new sense, as a predicate in superposition with science, generic rather than philosophical. Christ is not under Christian condition; it is rather Christianity that is under condition of Christ. We must subtract him from the involuntary founder of religion that he has become.

2. *Their unilateral complementarity as a disentanglement and new entanglement.* A subject is generally required as an operator, but one that often (not always) is effaced in the universal subject, and thus maintains the role played by "we philosophers" in the idealists (Fichte or Hegel). It is inevitable that the lived operator belonging to the context should be attached

to what amounts to an experimentation (the function performed by the evangelist disciples) and contribute to its materiality. But it is the two quantum principles, superposition and noncommutativity, that produce the general form of generic subjectivity, and in this sense these principles are subjective principles in the generic sense, without ever being subjects in the philosophical sense.

AGAINST "CHRISTIAN SCIENCE," "JEWISH SCIENCE," "PROLETARIAN SCIENCE," AND SO ON

There is therefore no subject-of-science but only a generically lived subject-science completed by an occasional subject. Whence the concept of what must indeed be called (cautiously) "non-Christian science," and which Christ practiced in a precise sense. On the one hand it is a science of religions—of Judaism, paganism (polytheism), and Christianity itself (to the extent that it has become a constituted religion). There is no Christian science save in a religious or ideological sense, no more than there is a proletarian or Jewish science in the sense in which those words have been used in recent history—a science of any object whatsoever, but ideologically marked by its ethnic and psychological origins. To confuse the object of a generic science—the world and its subject—with a particular theological and philosophical belief, and to pursue that confusion in the determination of a subject in the psychological and positive sense, is a form of theoretical "racism." We must accept that those confusions are of a philosophical order, and that only the generic point of view is capable of enduring the inclusion of a lived that is neutralized, depsychologized, or deethnologized. It is because it is generic that gnosis avoids the confusions that customarily appear in the form of so-called Jewish science, proletarian science, or Christian science that we are burdened with. But there is a non-Christian and non-Jewish science in the sense that within them the religious determination, the religious predicate, is submitted to the scientific cause or procedure, which imposes upon it the generic "non-" rather than its inverse. A Christian, Jewish, Greek, or Islamic ego does not belong to the science of religions itself. If the religious lived intervenes there, it is an entirely deindividualized one. By virtue of this generic lived, a science of religions cannot be purely objective or positive, nor given over to the phantasms and resentments of one party or another. The idempotent insertion of a generic subject into scientific procedures is the only radical means for eliminating positivity from the sciences without handing them over to a transcendental subject.

The predicate "Christian," or "proletarian" and the like, does not, therefore, touch on science save through the extraction of its subject, through the unspecified lived that may be that of a man of faith, a militant, or a lover. It does not touch on science itself, for there is nothing in science that can immediately be said to be bourgeois, Jewish, or proletarian. These predicates simply cannot be said, by way of some philosophical return or appearance, of a science that has nothing to do with any of this; they can be said only of the subject, and the subject is destined to become generic through its superposition with science, and thus to become nonproletarian, non-Christian, and so on. So it comes to pass that there is a subject that originates as Christian or faithful, but that is generic and nonindividual, a proletarian lived issuing from the anticapitalist struggle, but generic and not at all individual or collective, a playful and freely enjoying subject of art, but generic rather than individual. The subject and its directly worldly predicates complete the generic constant, but they are no longer the individual agent or the technician of philosophy that double the conceptual persona.

Seven

THE TWO LAWS OF SUBSTANTIAL RELIGIOUS EXISTENCE, AND CHRIST AS MEDIATE-WITHOUT-MEDIATION

THE INVENTION OF CHRIST

The invention named "Christ" was not the foundation of a new religion to rival others; it was, rather, a form of corruption that came to augment and complicate the imaginary grounds of the "three monotheisms." It was to determine, at last, the conditions of a somewhat rigorous cognizance of religions. Judaism and Greek thought utterly lacked the means to establish such a rigorous science of religions—an excess of monotheism and an excess of polytheism being their transcendent principles. These are religions that we shall call substantial, which are endowed, by way of religious content of multiple (possibly quite ancient) origins, with a form that Christianity, particularly plastic and adaptable, tends to avoid—not, of course, without certain tensions, since it too wound up as a vehicle for many old memories. We might say that Christianity has little or no substance, that its theological substance is a secondary spiritual production, or one that came after the fact. Torah and Logos are both too substantial and lack a relation of affinity in principle with modern science—they have no more than a circumstantial or occasional relation with it, they may have sought to engender it more or less easily but remained definitely severed from it in their principles. Christianity, for its part, supposes and yet goes on to dissimulate once again the discovery of a nondivine constant, which was not the case with other monotheisms or polytheisms, since substantial religion is precisely the refusal of radical discontinuity and the culture of an originary continuity with a mythological or even historico-political ground. The twofold religious material of Christianity, the two Laws from which it came, may give the appearance, let's say, of a synthetic path, with the idea of a completion sketched out on the horizon. But since we sense in it a greater affinity with the possibility of a science that is neither Greek nor Judaic in its principles, we must ask why Christ is defined as Mediate-without-mediation, and snatch him from his putative parents, who abusively lay claim to that filiation. The christic messianity of every human is the new "foundation" of a knowledge that refuses the substantiality of the ancients. Our faith-without-belief is that of an atheism that makes use of Christianity, but within the limits of science.

MODERN KNOWLEDGES AND UNLEARNED KNOWING

The genius of a founder of science consists in finding a scientific *stance*, rather than a philosophical position, in correlating the objective constant that ensures an exit from the circle of the old religions, and the specific lived that it subjectivates: the scientific procedure and the subject that, in this case, is called faith. Between the evangelical-worldly Jesus and Christ as a type of "Christian" stance, there lies the corpus of fundamental sayings that the former enunciated in principle. But Jesus himself produced a rigorous, if not exact, theory of his stance. He is therefore not only the spontaneous inventor of this stance but its theorist as well. In this sense he is, for us, the way and the truth.

Three great differences, nonetheless, distinguish this christic knowledge related by the Gospels from what positive modernity and even theology call the spirit or orientation of "science." The first is that the Gospels are clearly not the practical or epistemological treatment of a positive science, the reports or laboratory notes with which the hard and mathematized sciences have made us familiar. But nor are they a system of theology designed to crown faith, or a system of philosophy destined to ground it. They are what we call a generic science or a unified theory. The Gospels relate a lived experience of the world that has all the characteristics of an experiment concentrated on the Cross and the Tomb, but that has no meaning save through and in the Resurrection and the Ascension.

They are therefore not a discourse that stands at a distance from its object. The Gospels are performative or subjectively experimental; in them, the subject is not severed from its object that stands before it, but stands in a relation of unilateral complementarity with this object, between a status of generic condition and a status of operator-observer. It is engaged twice, and in a unilateral or complementary manner, in what we call a generic stance that *conjugates* a science and a philosophy that are entangled with each other.

The second difference is that here a new distinction, gnostic in spirit, passes between the knowledge that one is or that one practices as a knowledge that is immanent to faith without knowing it, and the explicit and scientific cognizance that is produced by means of theoretical instruments that are put to work in the same faith, through their resumption or their reemployment. It is the gnostic distinction between the faith that is already, like messianity, a knowledge, but a secret and clandestine, unlearned one, and its explicit or theoretical resumption (rather than its repetition). For us, however, the knowledge that one is without having it is truly what it is, or would be, for modern nonreligious gnostics: the knowledge of the world such as it is given by the entirety of modern positive scientific knowledge, arts, and even theology. But the positive knowledges that are, at first sight, particularly "learned" are, on second glance, seen by us as forming an unlearned knowledge, a tradition which is that of modern knowledge in the sense that it knows nothing of its generic sense, leaving us with nothing but oppressive or harassing force, and not "human cognizance" at all.

This difference is overdetermined by a third: the procedural, experimental, and theoretical character that gnosis must possess when it is not overshadowed by religion. This distinction is no longer that of modern knowledge, but is instead sedimented in a tradition and its generic cognizance; it is that of a state vector as a preparation of the Christ-system, of its probability amplitude and its probability of presence (to take this vocabulary up once again), from its prepared state to the state where it is detected as present and actual by the theoretical cognizance of the faithful or the nontheological.

Is it a matter of working "within" philosophy and theology, of raising Christ's foundational or inaugural formulae to the level of the concept? This return to the authority of the concept is prohibited for us by the very definition of our project, which is not to philosophize Christ yet again, but to philosophize "in" the science-according-to-Christ. We work with concepts, of course—it is impossible to do otherwise—but in the service of another thought entirely, reformulating the content of Jesus's sayings by conceptual *means*, above all in regard to a certain Greek side of Jesus or of his *Christianization*. But these concepts themselves are underdetermined and placed in another "position" or "orientation," given another function, precisely that of a scientific discourse, something we are authorized to do since from the outset we have exited from the game of hermeneutic constraints and the circle of religions and philosophy. Thus we neither thematize nor elucidate the meaning supposedly hidden in his words, unless it is to bring to theoretical discovery something that is simply implicit in them and already available to the simple—for Christ says everything clearly, with a self-evidence that is not accessible to worldly scholars but only to those who we call the "idemscient." We do not see in them *ultimately* any philosophical or dogmatic meaning, as if we read the Gospels in a foundational or even fundamentalist manner; we read them, rather, against every philosophical and evangelical fundamentalism. We identify, in the form of subjective axioms or lived oraxioms (axioms in-the-last-instance), the nonphilosophical coherence of the popular logia, so easily captured by philosophy, which loves to take charge of popular sayings so as to "raise" them up. With those simple (not philosophical) words, Jesus became Christ without having any need of the Pauline dialectic; he expressed his "identification" with this exit from the religious, that of Christ "in" himself. What could it mean to speak of "philosophizing in Christ" if he is not the condition that determines his own words and subtracts them from Judaism no less than from Greek thought? He is the scientific-immanent cause, the generic cause of the kerygmatic sense of his words; everything else is the effect of a Christianization that took place after the fact.

THE INTERPRETATION OF MEANING AND THE MEANING OF INTERPRETATION

The rigorous interpretation of the words of Christ as the propositions of a science, or as the oraxioms belonging to such a science, is an experiment that requires a quasi-experimental "preparation"—like the preparation for the "sacrifice" on the Cross. A formula of Christ's cannot be taken up like just any old discourse without being prepared, albeit under the general and "macroscopic" conditions of belief and theology. Such a formula is prepared, first of all, as a function of a certain device that enables the superposition of its interpretations or states, both Greek and Jewish, its only two possible contexts of reception and formulation. It is impossible to produce from the message of Christ and as a function of its local reception alone a meaning that is anything but blurred, suspended, or indeterminate and universal for all nations; this is what proselytism endeavors to do, all too happy to expand and protract Christ's message. But what is at stake is the actualization of the promise or the messianity that the message contains, in the form of a probability amplitude to carry it out or fulfill it, to bring it to a probable form of presence.

We wrongly project, retrospectively, onto the generic Christ, or Christ-in-person, the measure that we take of him, when we should be thinking of him as indeterminate. Religions, especially those that have a founder, are supposed to be contingent or historical phenomena that take their meaning from a history that is appended to them. But the history of interpretations of Christ simply constitutes a reservoir of variables that enter into his state vector. The probability of his coming remains indeterminate; he shall come, but we know not where or when—for the Greeks, for the Jews, and for the cultural and mixed subjects that we are. Paul supposes a Christ-event that has already intervened, a resurrection that has already taken place, from which he builds a truth to declare, or that is declared like an event to be announced, as something that has already come and that shall come again. His problem is that of identifying himself with the Christ who is already dead-and-resurrected, and of living through the proxy of his representation, rather than living him through superposition in the matrix of the Cross. The whole theme of the "return" of Christ in each of his faithful is affected by this temporal determinism. If Christ is our contemporary, he is not so through a projection of the past onto the present, unless we want to make his contemporaneity a Greek reminiscence. The return can be neither *from* history nor *into* a history. It is an undercoming, a sub-venience, the "evolution" of a state vector in the direction of a cognizance of salvation, rather than a reconfirmation of ecclesio-centrism. Christian theology and its religion are something like an inversion of the real phenomena through consciousness and representation—a negation of the future, which is to say, of the authentic meaning of Christ. Kierkegaard justly accuses the dialectic of conceiving contraries only in terms of their past or future, of having already surveyed them from the perspective of eternal time, and of not confronting the experience of the arrival of contraries. For Kierkegaard, experience is transcendence, but for us it is the most simple transcending: that of radical immanence.

THE INTELLIGIBILITY OF THE UNINTELLIGIBLE KERYGMA

How else is Christ at least real, if not possible? By way of what oraxioms could we "posit" him as a secret immediately destructible by philosophy and theology? Christ's message possesses an unintelligible aspect; it is, at the very least, a *paradoxical fact* according to the interpretation that Kierkegaard gave, still within the limits of his dialectic, which was already the most well-formed, most complex instrument of classical rationalism. It was therefore more than a datum or the factum of an existent reason. Perhaps we must forge a new simple (if grammatically difficult) term and say that *Christ is the* futurum *of a kerygmatic reason that is still theologically inexistent, because it is essentially quantum in nature, and must be decided or invented each time anew on the basis of the oraxioms contained in the message.* We must carefully distinguish the message from the kerygma. We cannot understand, beyond the message and its unintelligible intelligibility, what is announced, by way of an understanding (even a conceptual and dialectical understanding) of its dogmatic and past generalities. So we must grasp its kerygmatic amplitude as futural, sub-venient, not just as historical or as having taken place.

The paradoxical or unintelligible character is not, in general, primary; what is usually primary in the message is that it is comprehensible in the usual and more or less immediate, or even macroscopic, way. It is therefore an objective appearance that has its truth, but that should be generated from or within the compass of an experiment that brings another temporality into play. It must be admitted that historical causality exists, but what does not exist is a causal and linear determinism in which one may presume to inscribe the classical history of Christianity and the events that the message gave rise to and upon which Paul still grounded his interpretation. This interpretation is an objective, representative, or "decoherent" appearance that extends the primary unintelligibility of the message instead of yielding a nonunintelligible cognizance of this unintelligibility. No doubt it is, as the atheists say, a fable; but it should be rendered intelligible *as* an incomprehensible fable, or rather as a theoretical miracle. A science of miracles—*that* is what the science of Christ should be: a science that should invent the kind of intellection that does not rationally deny the miracle, but rather clarifies it in the only scientific manner adequate to it. The Christ-event is a theoretical miracle, which we are asked not to believe in but to actualize in faith. Belief is the basis of Paul's classical Christian interpretation, but it is nothing but a belief. It is not the faith that is science itself and that is unbelievable in terms of the codes of belief and classical reason.

The non-Christian science may look like an idealism of language grounded in a christic realism. But the situation is quite different. It is a "non-"conceptual materiel formalism, which uses straightforwardly algebraic quantum-theoretical axioms, but uses them as oraxioms to define a science-subject. If there is a problem close to the one we evoke here, a problem of "empirical" interpretation, it is settled by treating the Gospels as a modelizing of this theory, and nothing more. But theology precisely refuses to take the Gospels as a mere religious modelizing of christic science. The counterproof of the oraxioms used earlier for idempotence and superposition is that a governing transcendence or transcendence in principle such as God converts the messianity/faith duality assumed by Christ as mediate-without-mediation into a macroscopic triad, thereby causing it to lose its proper consistency, destroying the vectoriell superposition of its axioms.

THE SCIENTIFIC DEFICIENCY OF THEOLOGIES

Polytheism and monotheism, Logos and Torah, have each contributed to the formation of something like the "scientific mind." No defense of Logos's right to this claim need be mounted, and, for its part, the Torah, with the invention of the studious Jew, has traced the outlines of modern science, or intersected it without, for all that, clearly penetrating it. The Most-High-Other is a mask for a science that asks only to be made manifest; science must come to be—this is the scientific side of the Law that a Jew must never refuse. This is what all of science owes to Judaism: the Law as constant, the Law as text, the Law as obedience. Almost the entirety of the scientific milieu is indebted to these three. Despite its historical conjugation with Logos, the science of God cannot be theology, which forms a vicious circle with its object and fails to satisfy the necessary conditions that depend on an appearance—conditions that are other than those of "philosophical science." But theology serves as a material for us. God, from this point of view, is the set of discourses about him, about his power, his creation, his plan of salvation, his sacrifice, and his incomprehensibility, in all their theological variations. God as an object of science is the set of religions taken as the experimental field of his properties. This is to say that a somewhat rigorous and contemporary science of the Christ is to be built on the ruins of theology's "scientific" pretensions.

In any case, the refusal of philosophical and theological representation is not a refusal of all logically organized discourse. Christian science is expressed by oraxioms (which utilize concepts while requalifying them as primary terms, precisely through a suspense or neutralization of their philosophical meaning, thus repeating the christic operation on the level of the word), a set of oraxiomatic but immanent statements, infinitely deployed as the phases of a flux. They are neither especially cataphatic nor apophatic, and they undo theology, whether positive or negative; they are the generic phases of messianity, of the messianic "wave function."

However close philosophies may be to their theological offspring, a fundamental reason prevents them from having any status as *contemporary to Christ*. They are definitively disqualified by the quantum-theoretic procedure, which replaces the rationalist and determinist one, and which consigns them to antiquity and modernity, which is to say, to philosophical finalities—a fate that is avoided by the consideration of contemporary theoretical means, taken as simple productive forces with only a generic finality.

So it is that the kerygma of Christ, as given in his affirmations and exhortations, shall no longer be taken, in traditional fashion, as an object that could be submitted to a historian or an exegete. It will be submitted, rather, to a quantum-style experimentation. We shall therefore suppose that the kerygma comprises superposed states of language, of only apparently identifiable origin, and that it can be inserted into a quantum-style interpretative system or experimental apparatus, after the model of the so-called two slit experiment. In its output can be read either a Greek immanence or a Judaic transcendence, depending on the apparatus of measurement or output selected (the Churches), which can make one or the other appear. Thus the historian or exegete will define a Jewish Christ or a Greek Christ depending on which apparatus is selected—that is, depending on their criteria of interpretation. This is simply a vicious circle. We cannot deduce from this result an original definition of the kerygma as being in itself, really, either Greek or Jewish (*Either/Or*), or even as being a mélange of the two that could be analyzed and synthesized by a historian or an exegete. This possibility is now excluded, and we cannot imagine that there was formerly a mélange of the two, a hesitant combination. The message has no meaning in itself, to be recovered precisely in the form of a definition or a reproduction of the criteria of understanding; to assume this would be to neglect the transformative work of science. The kerygma of Christ, which is certainly indivisible, is not divided between Greek and Jewish origins, but distributed over the whole assemblage, which includes both the emitter and the receptor subject, and thus bears the mark of the apparatus's finitude.

This allows us to combat dogmatism. The Churches, in particular, are in a position to dogmatically capture the message and attempt to appropriate it as the sole authentic truth and to make it an in-itself. This supposed in-itself is like the dogmatic or natural attitude of which Husserl spoke. This projection of the reception or the measure onto the supposed in-itself of the message is an essentially religious scientific error, an interpretation that confuses the experimental vessel of the kerygma with history as presence. One cannot say in a determined and assured fashion who Christ was or what he meant to say. His formulae, such as "children of God" and so on, are intelligible in the first degree, for common or theological sense, but are thereby deprived of their pure kerygmatic character, which can be reached only through faith. The kerygma is the complex, indeterminate, or "fictional" sense of the message, that which is vectoriell in it. If there is a problem of interpretation, it must be understood to be one that unfolds in objective appearance and that is religious-philosophical. The kerygma is generically vectoriell, it sub-venes in a radically immanent fashion, or abandons double transcendence through the algebraic "quarter-turn" as immanent messianic opening.

THE SYMPTOMS OF THE MESSIANIC STANCE

What did Jesus, qua Christ, say? There are symptoms in the message of Christ, religious data necessary for the preparation of his quantum-theoretical treatment. This was not, as such, achieved in the Gospels, but we shall base our work on these symptoms in order to chance such an experiment. For the exegetes, he describes his mission in two different languages. Each time it is, apparently, an implicit identification, more practical than theoretical, in accordance with a Law: the Torah, of course, but also that other formal structure which is the Greek or philosophical Law, the Logos, which is perhaps more implicit, given the Jewish environment and origin of Jesus, but which is just as fundamental as the Torah, and which is too often forgotten when taking stock of the constitution of the christic achievement independently of the Greek Fathers. Pagan Logos and Jewish Torah are substantial religions, even if they can also be taken as forms of other cultural modalities. They are the forms of existence that Christianity will go on to inherit and with which it will mix itself, as if to bypass Christ and escape his radicality. They are consistent, forming a system or a plan of salvation, and they are structured according to three terms: the divine, the human, and the law that serves as an intermediary, whether this be the Logos as mediation or the Torah, which is another mediation in which the three terms enter into other relations. But the problem of Christ is, perhaps, radically simple, since it is precisely the problem of the Simple Ones, that of making a nonreligion with two terms out of the three of religions, simplifying them in their essence and their relations. In other words, Christ's problem is not Pauline, still less Hegelian; it excludes every attempt to solve it by way of a triad of instances or a trinity.

On the one hand, Christ describes his mission in terms that evoke a mediation, to speak in Greek or philosophizable terms, a weak or distant mediation between God and men, crushed and as if having lost his divine side, which he does not "represent." He does not re-present, a second time, his Father, but rigorously presents him; he says that it suffices to have seen him to have seen the Father. In his person, he fulfills this mediation, but on condition of suppressing the infinite, and a fortiori the finite distance from God (polytheism), short-circuiting this mediation, as if the middle term contracted the entire set of terms and took them upon itself. This fulfillment of a contracted mediation transforms the other (human) side, assures the salvation of men, of whom a new form of obedience is suddenly demanded. This is not, therefore, the developed mediation that Hegel would later see in him, but a mediate-without-mediation, which is to say, without a complete or triangular structure.

But he also says that he has come to fulfill the Law, to speak now in Hebrew—that he is this fulfillment of the Law in person. We can also understand this formula in Greek, because the Mosaic Law has certain appearances in common with mediation. But just as we earlier philosophized in Christ, we now Judaize in Christ, and in order to do so we have to have already Judaized as radically as possible. This mediation is altogether different from the Greek one. It is at once an immediation of subjects to the transcendence of the Law, by which the Law demands the greatest, most precise, maniacal obedience of humans, and also the greatest distance, because it testifies to the infinite separation whence God affects us with, precisely, the Law. An infinite proximity of obedience to the Law, but a proximity that is the same thing as an infinite distance from God.

There is therefore a double postulation of religious language with Jesus, and both times a fundamental structural modification is introduced, one that comes down to fulfilling in his person, and therefore assuming in an altogether novel fashion, the substantial Laws, which he partially empties of doctrinal and theological content. Christ is the concentrate or the contraction, the conjugation rather, of both Laws, because he is the one who generically incarnates them and who profoundly transforms them with respect to their forms of authority. An exclusively Greek intellect here risks straying toward Paul and the dialectic, while an exclusively Jewish one risks straying toward the prophets; but the two Laws alluded to have a general function or "common" aspect of mediation between God and men, even if one tends toward the automediation of God or Being and the other tend towards hetero-mediation, if we may say so, by and as the Other. It is once again necessary to put the proposed procedure to work, so as to verify what has been presented in this summary description.

THE PROCEDURES OF A RIGOROUS CHRISTO-FICTION

Messianic subjects, or the faithful, may also be called the "idemscient," an idiomatic term that could be applied to the gnostics also. From the outset, they have knowledge of idempotent superposition and noncommutativity in general, the two principles of quantum-theoretical thought that govern the relations between Christ's words in their immanence and the transcendent statements of theology. They allow us to obtain the oraxioms that describe and constitute christic science, which "prepare" it, as physicists say, for experimental measurement.

1. One must "superpose," in the quantum-theoretical sense of the word, on the one hand, the logia, all the statements on the Law and its fulfillment, on the relations between the Father and the Son, and Jesus's countless formulas on life and the living, and on the other the Greco-Judaic theology that is a transcendent interpretation of all of this. The "addition" of the messianity of christic words and theology should (through transformed or axiomatized theological material, having lost the burden of sufficient theology) restore the messianic force of the christic sayings. Theological discourse is transformed, its internal philosophical structure bending to the principle of idempotence.

2. It is a matter of eliminating not only a speculative sublation between these two types of statements, but also the simple reciprocal and specular imitation of commentary—gnosis is more than a gloss. The time and space of messianity do not belong to the same world, or even the same history, as those of theology, because messianity is not mythology extended into rational history. Christ produces the ultimate instance without producing a norm, since as Last Instance he remains the same ultimatum without allowing himself to be affected by theological discourse. The only transformation that takes place here is that of Judeo-Greek speech, as it is rendered adequate to the person of Christ and loses its sufficient theological transcendence.

3. There is no commutativity between the words of Christ and Judeo-Greek theology, between belief and the cognizance of faith. This principle endows messianic immanence with its being-foreclosed, and its a priori defense against philosophy and its "corpuscular" or "atomistic" representation of the real. We must understand the words of Christ in such a

way that there is no reversibility between their effect and the theology that is sutured to them. The "return" of Christ himself in the form of his theological intellection, since philosophy demands only the circle (albeit infinite) of the real and language, is impossible. Every theology is either of the order of the philosophico-religious imaginary, a formation that serves us as material, or else, if we elaborate an axiomatic principally according to Christ, with this imaginary as occasional material, it will no longer have any constitutive grasp of messianity but, rather, the latter will express itself through it and give rise to a new Gospel. Our science can therefore be called a *rigorous christo-fiction*, for it is no longer imaginary like the material from which it is fashioned. Theology, and Christianity in general, are thus engaged in a transformation that is deeper than a reform or even a "turn," the last residue of the old circularity. This is what we call the unilateralization of Christianity, or even the a priori defense of Christ against Christianity.

FOLLY AND SCANDAL

We refuse the traditional theo-christic or theo-anthropological couple of the figures of Christ, welded together by the mechanism of mediatization or schematization. Christ is neither a simple historical datum reduced to the problematic figure of Jesus, nor the founding factum of a relatively rational belief, of a Church faith elaborated by Christian theologies. He is rather a "scientific given," but one of a new type, a generic stance rather than an empirico-rational "fact," but a generic constant that places its two main types of interpretation under an underdetermining condition, single-handedly transforming them into the statements of a science. We can always imagine a transcendental Christ, or rather a Christ reelaborated as existential; but what we seek are his vectoriell and immanent determinations, which make him underdetermine the Churches' Principle of Sufficient Theology.

Judaism and Greco-paganism are routinely called upon by theology in order to analytically and/or synthetically (dialectically) comprehend the christic stance, or to "make sense of it." These are the two extreme law-materials: Greek, on the one hand—the Logos strictly reduced to its own kind of immanence, the principle of the Eternal Return of the Same, for if there is a polytheist religious principle, it is indeed that of the Vicious Circle, there are no others. And on the other hand the Torah, which infinitely defers the coming of the Messiah. But Christ is unintelligible for both of these laws—law as vicious circle, law as infinite obedience—each of which, in their own way, decomposes or deconstructs the indivisible stance of Christ only to recompose it. We could give this an exclusively Greek reading, on the basis of the externality of the Son to God and of the Son to man, a reading in terms of mediation, therefore, with all the avatars and narratives that have accompanied the history of mediation up to and including Hegel. Or yet another reading, a more Judaic one this time, as a concession to the Greek, for which it is not a question of mediation but of the practical fulfillment of the Law, here again with all the interpretations that accompany the notion of the Law's fulfillment, from Paul to Hegel. In presuming to render it intelligible through their combination, Christianity, which thereby believes that it has made some progress, falls victim to an "immanental" appearance, which causes it to confuse mélange with scientific superposition. Christianity ceaselessly murders or destroys Christ, whether by placing a transcendent messiah in a Jewish sort of waiting, or else with a philosophy of the messiah's "return." It should rather be said that Christ is the superposition or quantum-addition of the two laws, undecomposable into identifiable or measurable factors. How can we submit these two problematics (which are conjoined and entangled in the mélanges that, for theologians, form the fabric of Christianity) to the fundamental principles that are made use of in quantum physics, which serves as our guide or model; and through what transformations must those problematics pass in order to rejoin these principles? From these interpretations, which served as the bait for, or gave rise to, so many theological variations, superposition and noncommutativity strip from the first its Greek trait of the milieu or intermediary, and, from the second, its Judaic trait of extreme transcendence.

Christ does not propose another law that revises the Greek and Judaic laws, but instead a law-event, a generic law, which is that of superposition and noncommutativity. Christ is the law fulfilled, as the immanence of messianity, as *idemmanent*. This is not a sublation of the law, but its suspension and depotentialization, insofar as it belongs, through idempotence, to the Mediate, to the Mid-site, the milieu, of analysis and synthesis. Christ realizes or fulfills them (as materials, therefore) by "identifying" them, no longer with each other, but with an altogether other law, also double, but in another way: the law of the idempotence or the generic and messianic nature of his mission, and the law of the non-commutativity of that mission with existing religions. A victim of religious and philosophical appearance, Christianity will thus have grasped that the "fulfilled" law is still the Jewish Law or one of its modalities. But the new practice is foreclosed to the Jewish Law and the Greek Law, and can therefore transform them discontinuously into an altogether other principle, which is that of messianity. In what does the "folly of the Cross" and/or its "scandal" consist? Most certainly the fact that, in Christ, traditional opposites enter into collision. But Paul offers man a still rather classical combination, which bears the visible and ideal, corpuscular trace of the old materials, the quasi-dialectical mélange of an already too Greek and contemplative law of universal love, and an event of the empty tomb, which idealizes the Crucifixion and the Resurrection on an abstract basis. Paul is a "converted Jew," and therefore a paradox and a contradiction in his own right. His doctrine is itself a "folly" from the point of view of rigorous thought: it is a coalition of determinations that will give rise to multiple interpretations. His universality is too "easy" for the Jews: the law of love, far from being generic, is a practical law—a law that is not given as an ideal, but that must be received and exercised as an ultimatum, in the act that takes the world as an object. Its function of mediation will even be understood in a Greek fashion, allowing for the reconciliation of all men in and through a universal mediation. But if Paul is the founder of universalism and of that intolerant abstraction that is the Church, then Christ is the inventor of the generic. He is the law as messianity, the law as faith and not as belief, the law of generic love, which is not indifferent confusion or the equality of universalism. In the two unfavorable interpretations, thought remains with a transcendent God delivered to his arbitrariness, grounding a generality of the law as jealous tyranny or even as universal love. Mediation must be reduced in a new way, or else it will remain automediation and will not be unilateral, and the fulfilled Law will remain mediatized and sublated, divided between objective belief in transcendence and sentimental, customary, or obedient faith.

THE KERYGMA AS SUPERPOSITION OF THE CONTEXTS OF THE "CHRIST" MATRIX

We shall not say that Christ is, in himself, a machine, though he is indeed one in certain senses—for example, insofar as his various moments are determined by the indivisible matrix that he forms. Perhaps we could speak of an algebraic machine. He is, at least, a matrix that functions through the addition and superposition of contexts of reception or interpretation.

Consider any one of Christ's sayings under the form resulting from the scrambling of two interpretations, each containing its various possible properties, including certain mixed or ambiguous theoretical determinations. For the moment, it is nothing but information, and still not faith. We then enter it into the matrix or quantum apparatus formed of two channels that, in Christian theology, simplifying matters a great deal, are "Logos" and "Torah." Coming out of this apparatus, this saying can be taken in a clearly defined either Greek or Jewish sense, as if the apparatus of reception had served to sort out those determinations. The chosen or detected result in such conditions strictly depends on the explicitly chosen medium of reception or detection channel. If the theoretical apparatus is Greek, one will obtain a formula with an obviously Greek and philo-sophically interpretable meaning. If the reception is conducted by a Judaic apparatus, one is then referred to an altogether other context to find the meaning of the formula. The same christic formula is susceptible to two heterogeneous meanings, two interpretations with no common measure, and apt to nourish wars between the religions that try to capture the mes-sage. This is, in general, what happens to theologies and even to exegeses of Christ's message that spontaneously place themselves a priori at the output of their own interpretative apparatus and try, in their respective fashions, to clarify the equivocations, to find their way through a whole mess of amphibologies, analogies, and mélanges of traditions and influ-ences, back to the unitary and univocal meaning of Christ's words. They are involved only in the sterile reception of what has already been placed in the apparatus.

Those of us who forge a nondogmatic experience of faith, and who prepare it in a quasi-experimental fashion, do not have to place ourselves at the level of what is already an "output," or a predetermined interpreta-tion, folding it back upon itself to form a dogma, a simple duplication of information, but not a cognizance—strictly speaking, creating an item of theological knowledge but not a faith. In reality, we should give up on deducing, a priori, that which was sent, the conditions of messianity and its cognizance, on the basis of its channel of reception, falsifying a priori the effective meaning of the message by only considering the apparatus of detection and its a priori projection onto the vectoriell givens, and thereby presuming a knowledge of what has actually taken place in the Christ-matrix. *We cannot draw any conclusions from historico-theological symptoms sent to us by religions or confessions, pertaining to faith as "authentic" cognizance.* The theological states injected into the apparatus are no longer treated individually, each on their own behalf or in their mélange considered as an all, submitted to either an step-by-step physical causality or a metaphysical causality, in the sense of the creation and models of that type, and hence in a determinist spirit, mobilizing one of the "Aristotelian" forms of causality through which one can continuously follow the work of elaborating the data. The data are treated as vectors and superposed or added to one another as vectors. The latter cannot therefore recover either the input or the output data, or the results, which are macroscopic in every way and correspond to a destruction of the internal quantum operation. The outputs are measured through the output channel but are not measured "for" that channel—this stands for faith and its messianic content, for which Logos and Torah are the variables or productive forces but certainly not the ends or criteria. For the well-defined output channels are no doubt necessary, but are not sufficient for really determining the meaning of what is ipso facto detected in the matrix, which is to say, "in-Christ." They function as interfering channels, whereas, in the output, they are themselves isolated and taken up again macroscopically. The great theoretical rule is to explain that we must be ready for what happens in the cognizance of Christ, because of or in virtue of the fact that we are transformed by the received faith of Christ. We shall come back to this complementarity of messianity and faith, which forms a unique vector, conjugating, in faith, the prescience that is generic messianity and individual decision.

The cognizance of messianity—that is to say, faith—is distinct from the knowledges injected into the matrix, knowledges that we presume to form a linear and deterministic continuum of causality with the output channel. Theological knowledge does not definitively yield any cognizance of the kerygmatic essence of Christ. Macroscopic or doctrinal interpretations, be they Greek or Jewish, through their continuity destroy this essence, which carries various vectors through addition or superposition, and which forbids us from fixing, a priori, what really takes place in the black box that is Christ at work, or what the essence or the real of Christ might be—something we only come to through the effects of faith, as the probable cognizance of messianity. It is therefore a question of rendering intelligible, even if in the form of a nonknowledge internal to science, the internal and properly quantum trajectory of the superposition of the state vectors of the christic message, and hence its detection as messianity and as faith. Just as the superposition of state vectors cannot be confused with the simple data that are still not vectoriellized since they lack precisely the vector-form or complex number, so the detection of what the vectors have become cannot be confused with the "corpuscular" results that repeat certain more or less theological variations. Superposition and messianic detection are original phenomena that have no common measure with a brute, experimental science of faith or even with a spiritualist theology. They are operations hidden in the hearts of the faithful, precisely the unintelligible or unlearned knowledge hidden from the philosophers and theologians that are the eyes of the world.

THE FULFILLMENT OF THE TWO LAWS

Let us draw out the structural implications of the result obtained from the output of the apparatus. We shall leave aside for now the fact that this output is, if not aleatory, at least marked by theological probability, and shall call it "faith" as underdetermined by messianity. The fulfillment of the Mosaic Law to which Christ, in his person, lays claim is a way of transforming it, and the Greek Law with it, by assuming it. How does this fulfillment take place? By suppressing this time the proximity of the Torah, just as, before, the Greek Law was fulfilled through the suppression of transcendence. He liberates men, this time, from the infinite, crushing proximity of the Law, and, in order to do so, replaces it with the faith that leaves men free to choose between the absolute obstinacy of the subjugating Law and free obedience to Christ.

At the heart of all this is the immanent fulfillment, the transformation bearing not, at first, upon the infinitude but upon the transcendence of God and the absolute proximity of the Law that demands obedience. The mediation of three terms becomes a duality. The Law no longer divides me as a subject, as Christianity and, for example, psychoanalysis (Lacan) would have it, but distributes the entity "Jesus" in another way, as a generic or (ascending-) man without transcendence or messianic Last Instance *and* as a subject determined in-the-last-instance. It is the obedient ascension of the faithful. The Christ-Law is generic; it does not divide but distributes and dispenses, within messianic immanence itself, the immanence of faith and the transcendence of the Law—a transcendence that is, moreover, liberated insofar as it is now submitted to faith. The fulfillment of the Law by its being placed under christic condition is not its negation, its abrogation, or its sublation; it is the creation of faith and, simultaneously, the transformation of every man into a messiah. And the messiah is the stripping-bare of man of all the predicates that situate him in the world and impose upon him his coordinates, trajectories, theological positions, and relations, his being as a sinner and his becoming redeemed. Such is generic universality: *in-the-last-instance* he is no longer either man or woman, neither master nor servant—neither Torah nor Word.

IN-THE-LAST-INSTANCE THERE IS NEITHER TORAH NOR WORD

We can arrive at the same result in a quicker and above all more positive fashion by showing that the theory named "Christ" results from a superposition by idempotence of the two Laws or two forms of mediation, grounded respectively in the double transcendence that supports the Torah and the Word. Superposition through idempotence has strict conditions of application or use that are not those of a pure and simple identification, which would present that blurry image of Christ presented by Christian solutions, Greco-Judaic mélanges. In order to be superposed, the two Laws, with their information content on Christ, must be reduced to phenomena of vectoriellity, lest we go no further than the signified or the signifier (that is to say, no further than corpuscular languages) and relapse into the philosophical spirit of mélanges and doublets. A superposition produces, precisely, a generic image—not a mélange, but the generic form of the mediation that is Christ, a mediation that is neither Greek nor Jewish, but that, qua generic, can stand as an a priori for both religions. We have posited the "Christ-message" as a discursive system with two "states," or two classically discernible, perhaps dialectizable, properties, Logos and Torah. Now, their vectoriell addition (Logos + Torah) is *still a state of the message or one of the properties of Christ.* Christ remains Christ in each of these three states, no less so in the last of them. This is his generic universality, but with the decisive difference that he is thereby conceived as the superposition or addition, Logos + Torah, with neither of his properties being discernible or isolable. Conversely, if we were to try to identify them individually and separately, we would destroy the superposition that is always called "Christ." Christ is not a subject-point like God or the ego, an atom or a corpuscle that can be identified by the coordinates of a theological space; he is an interference, Greco-Judaic in origin, but one whose components are no longer identifiable. Superposition is not a synthesis, and least of all is it a mélange in which those components are simply scrambled together but relatively recoverable. On the contrary, any attempt to "unscramble" them is enough to destroy Christ, and lose that in which his indivisible messianic nature consists, which is no longer anything like a temporal or evental source-point, a history in a space, but something of an immanent and transfinite messianic flux. Logos-Christ and Torah-Christ, the Word and the Law, are henceforth only one in him, which is to say, in-One. They are indiscernible and occupy the entire space of messianity, which they ceaselessly open up. It is fundamental for faith to no longer consider Christ historically as deployed in the divided space of world-history, as an object of the Churches' belief. He is a radical interference who ceaselessly comes to us before the theological wall that he has already breached.

THE GREEK LAW AND ITS TRANSFORMATION: THE MEDIATE-WITHOUT-MEDIATION

What has happened? What has become of the common Law of the two discourses, which renders those two types of mediation immanent? Recall, for a moment, that superposition as the immanence of the Same has as its condition the suspension of an arithmetical operation of addition that can hardly be understood as numerical, and that posits its terms in a space of exteriority or transcendence. Two entities in a theological space are not superposable, while an idempotent addition produces a Same as wave and wave function. Idempotence is therefore immediately generic insofar as it programs the equivalence of a term with itself as a result, whatever the operation serving as mediation might be. Bracketing out the excess, it is the instrument of immanence through the suspension of every organizing and dominating transcendence. This indiscernible vectoriellity will, obviously, be the messianity that philosophy would rather decompose into atomic entities and external relations, or into substantial entities and internal relations, only to recompose them into a dialectical trajectory at the stroke of the operations of negation or affirmation, each time putting to death the unbearable indifference and devastating nonaction of the Messiah.

From the simultaneously Greek and Jewish side of God, there is a suppression, through his sacrifice, of his transcendence as distance, while the infinite is conserved and internalized in the immanence or "in-person" of Christ. And from the other side, that of obedience to the Law, there is a suppression not of the infinite but of the subjugating proximity to the Law, the contrary of transcendence. Christ is not located midway between the two Laws, between Logos and Torah, like Buridan's ass, like their in-between—the ass of a new Trinity, concentrating Law and man into neither a living Ego (Henry) nor a living Law (Derrida) but a new kind of duality. On the one hand this duality is an immanence, but one that is infinite; and on the other it is a rediscovered transcendence of man, but a weakened or abased transcendence entirely imbued with the infinite immanence of this new Law. Instead of mediation and triplicity, a new duality that neither divides nor synthesizes the human lived, but unilaterally distributes the latter between its immanence as the infinite Christ-Law in which it participates and its ancient transcendence, now rooted in christic immanence. This duality will henceforth, under the auspices of an im-mediation, unify that which will have been abridged or deprived of the doublet of divine transcendence and proximity, but which remains infinite as immanence. We must understand that the heart of the matter is immanent fulfillment, and that the modification bears not upon the infinite itself, but upon the transcendence of God and the absolute proximity of the Law that demands obedience. The mediation of three terms becomes a duality. In short, the Law does not divide me as a subject, as Christianity and, again, psychoanalysis (Lacan) would have it, but distributes "me" otherwise, without division, as the Man or Last Instance that "I" am and as subject determined in-the-last-instance or in obedient transcendence. The Christ-Law is generic, being neither Logos nor Torah, and it does not divide but concatenates its phases, first as vectoriell (wavelike) immanence, and then as (particulate) transcendence, within infinite immanence. Generic universality has the form of a unilateral duality, and not the form of a divided Whole returning to itself. In both cases mediation is transformed but not totally suppressed. Something of transcendence remains, but it is simple or unilateral—the transcendence of the obedient subject of proximity no longer playing its mirror-game with the divine.

On the side of transcendence, the divine infinite is sacrificed in favor of Christ. But the transformed Law remains the Law, now neither Greek nor Jewish; obedience remains the obedience of man who gains his "Christian freedom," a certain autonomy; it is an obedience determined in-the-last-instance as infinite obedience to immanent Law. Christ does not make us renounce the Law, but only its Jewish and Greek forms. He is its immanence, he fulfills it in his person and "at the same time" or rather "in the last instance" obeys it as undivided subject.

What, then, becomes of the mediation obtained and generically transformed, the New Law that results from it, but that is indiscernible, and that is Christ? The divine infinite is separated from its transcendence and returns to immanence, or is transfinitely contracted in idempotence. Christ, the Son, generic kernel of mediation, is immanent Same as quasi-infinite vectoriell flux. And this Same is the Law, which is destroyed when interpreted after the fact as Word or as Torah. The idempotent fulfillment of the Law is not its "suppression," but the generic way of "internalizing" the Law, or rather of unfolding it as immanent while conserving it as infinite. So there is a transformed obedience of the subject here. For, as we have said, Christ is the immanence of a duality without division, a unilateral or generic duality. This is, therefore, no longer the infinite obedience to the tyrannical that, in its proximity, is also transcendent, nor the Greek obedience to cosmic Reason, even as internalized into the subject; it is obedience to the transfinite immanence of the Same or the messianity called "Christ." The immanence of the Same has the form of a messianic flux, and fidelity is obedience to the messianity that guides the subject without either persecuting or commanding him, without turning him into a hostage or a subjugated subject. Instead, messianity is that which defends subjects against both Jewish and Greek forms of religious harassment.

"GOD MADE MAN": THE SACRIFICE OF MEDIATION

Take this formula, "God made man." What does "made" mean here? The belief that God has metamorphosed himself and decided to clothe himself in human form lies at the origin of the anthropomorphisms and trinitarian theologies of the Church, and of its aporias, to which intra-Christian heresies testify. But it is also the origin, to a large extent, of the modern dialectical philosophies or christologies at the margins of Christianity. It signifies that there is an immanent "becoming" of Christ as subject or, more rigorously, a sub-venience of the two religious laws into the genericity of science as lived—into the subject-science as true science and true subject. Christ is a scientific procedure that has found a lived use. Rather than becoming, the Messiah is coming; but he is not coming from the heavens or from the earth—he "under-comes" or sub-venes as generic, and therefore has no place in the world, precisely because for him the world is a complementary "object" to be transformed. To be "made," here, is neither emanation nor procession, neither their medium nor their synthesis; Christ (and, with him, each of the faithful) is said duelly (*se dit au duel*). The faithful is an unlocalizable duality grasped in immanence, the duality of a conjugation or a complementarity, a unilateral duality as a procedure of idempotence—that is, of the indivisible excess of the Same in the Same, on the one hand, and as the materiality of a lived, on the other.

Now, an idempotent term or operation, to quickly recap, remains the same, and is not modified as a result of adding or retracting the same term. What one may have been able to see as the transcendence of an arithmetical addition is suspended in its effects. Mediation is neutralized and ineffective, which does not mean that it fails to register, since it should be an idempotence and not an analytic identity. The lived logic of messianity is not the onto-logic that grounds logic in beings and that traditionally serves as a mediation in philosophies. Lived logic is generic and overflows the coupling that associates logic with beings, or even with being. Idempotence is, in general, the milieu or midsite, or the complementarity of the analytic and the synthetic—but what kind of "milieu" is it?

Idempotence can only be said of immanence, but it is a form capable of harboring a simple transcendence. Suspending its internal mediation, it is the nonmediatized and nonmediatizable element of Christ, the Same. But in the form of a Mediate-without-mediation it is also the element of . . . or for . . . transcendence, even though it is not itself mediatized with the latter. The generic is now complete: it is a Mediate-without-mediation = mid(site) = element, which mediatizes the other term, by which it is not mediatizable in turn. The generic Christ mediatizes-without-mediation the world, but is not himself mediated by the world. Here we will recognize the true nondialectical mediation effected by Christ.

At this level of principles, the milieu of the analytic and the synthetic is not an analytic identity. The generic mediates for the world but is not itself mediatizable. Thus, so that the nonmediatized mediate, the first A of A + A = A, can serve as a Mediate-for . . . or generic element for the other term, the former must necessarily be immanent. Idempotence and superposition comprehend each other, when they are lived, as a sacrifice of reciprocal mediation, of exchange with God, a unilateral sacrifice of mediation, on condition that God make himself generic man, the Son or element of the Milieu or Mid(site), the just, unequal, or unilateral measure. What is sacrificed is reciprocity or rivalry, the divine side of mediation, but on condition that God be radically immanent or himself nonmediatizable. But it is precisely immanence that is found or is realized by the suspension of his internal mediation of the two sides, and becomes an external unilateral mediation with the introduction of transcendence.

To disengage the kernel of mediation as unilateral implies that the messiah, as generic, is in becoming (which could stand for Judaic waiting . . .) and depends on the encounter-without-creation, on the "collision" that is Christ, the collision of God and the World and their superposition without synthesis. Messianity is in progress in the sense of its preemption over men, because the Christ-factor serves as an index for the human variable. Their addition suffices, as the intervention of God, not as creator of the world (the Messiah is not a new creation) but as Christ-factor of superposition, or as something that can be added. The Christ-event can be interpreted in terms of the concepts of superposition or quantum-theoretic addition, and therefore in terms of grace, but certainly not in terms of re-creation. It is not a question of a refund or an exchange, of his taking upon himself the sin of the world, but of bringing about the only genealogy possible, not a genealogy of the world but of its salvation, from and within radical humanity. Thus transformed, the new Sons or messiahs cannot be cognized or recognized for what they once were. They are messiah-existing-Strangers.

THE JUDAIC LAW AND ITS TRANSFORMATION: MESSIANIC NONCOMMUTATIVITY

We have already dealt with the two Laws and their messianic subvenience as Christ, but from the angle of the Greek Law insofar as it has a more obvious affinity, at first glance, with the principle of idempotence, of which it is the philosophical symptom. We shall now deal with the two Laws from the angle of the Judaic Law, insofar as it has more of an affinity with the second principle, the noncommutativity of the Same and the other, of which Judaic unilaterality is the religious symptom.

How does the sacrifice of God and gods condition messianic idempotence itself? We are going to exhibit the stance of the subject-science in the figure of Christ. Meanwhile, we have examined the question of this science's potential philosophical circle, and have settled it by showing that the subject-science is an emergent structure that owes nothing to philosophy, that Christ is an emergent event that owes nothing to the Greek and the Jew (except, perhaps, retroactively), and nothing, moreover, to its abusive "Christianization." For philosophy and religions try to capture the generic stance of truth, and thereby create an apparent circle. Noncommutativity is decisive for explaining the possibility of christic idempotence, and its immanental reduction to religions. It imposes a prior-to-priority character on this reduction, and protects Christ from every return to a first theology, governed once again by the transcendence of God, around whom it is organized. Christic science is not a canon but, strictly speaking, an organon for faith. Nothing, no order, precedes Christ, who is precisely not first but the "last instance" of salvation, the prior-to-first ultimatum. The genius of Christ—yes, the genius—is for the simple and is made to suit them; it is not "the genius of Christianity," which is, again, just a minor symptom. It consists in the intertwining of another principle with that of idempotence, the principle of a certain noncommutativity. As a principle, and not simply a law, noncommutativity amounts to placing irreversibility at the heart of the beginning and of priority (at the heart of first philosophy and its "principles"), rather than throughout a temporal sequence that would reencompass it in its circle. Inserted into priority, it doubles the latter, without creating a philosophical or specular doublet; it is a prior-to-priority that does not fall into an order so as to submit itself to it, but rather establishes an order for philosophies and religions themselves. Noncommutativity has a precise meaning: two axioms on Christ and his sub-venience combined in a sense that does not yield the same result as the reverse combination of those axioms. In a broad and qualitative sense, philosophical priority can only presume to commute with science at the price of any appearance of truth. The Christian stance is not first (it motivates no science), but prior-to-first. Truth, like philosophy itself, is a first thing, but the true-without-truth is prior-to-first. Philosophy and religions do not furnish truths in the strict sense, but dogmas, amphibolies, and objective appearances, sure enough: material to be dealt with. Instead of truths, what Churches have are Dogmas.

Noncommutativity determines messianity as a unifacial wave or a unilateral-oriented throwing. The New Law is in progress in each of us— we who are, in virtue of this fact, the messiah—and the messiah thereby sub-venes in the fulfillment of the Law in his person, not externally. He incarnates the Law of the Same in faith, which is not an external complement, but a redistribution of the manifest mélange of faith and belief. In this interpretation, faith is of the immanent lived, more generic than individual. It is foreclosed to the anonymous Law, which it reinscribes in the ledger of the world.

The unilaterality of the Judaic "Most-High-Other" now appears for what it really is: generically, a symptom or occasion of messianic noncommutativity or complementarity. The generic is the new status of the Law that underdetermines the obedient subject who it preempts (the generic concept of election) and structures, through its scientific but not positive nature. The Law whose proximity was the act of a God too distant to intervene among his subjects is concentrated and unilateralized between quantum-theoretic principles and the lived, impregnated by idempotence and rescued from philosophy and the pagan world. With the immanental reduction of the Torah, only this Law remains, and the prohibition is made (or, rather, the impossibility is signified) against religions and their theologies returning to Christ or interpreting him, he with whom they cannot be commuted or exchanged. The principle of idempotence allows the Jewish God (the One or the Other) to be superposed with subject-man, to realize a mediate-without-mediation (but certainly not an identification or a confusion) between the One-God and the human subject.

Linear superposition materially avoids a potential circle of messianity, and noncommutativity avoids it formally. Together they explain the wondrous emergence of the Christ-event and the creation of the science of religions that Christ ceaselessly formulates in his nonphilosophical fashion. It is obvious that he sets the being-foreclosed of his words, even those closest to philosophy, in opposition to a Greco-Judaic givenness of meaning. Noncommutativity must a priori protect or defend his popular or figurative formulas, and give them an axiomatic function. The intuitiveness proper to the vector of messianity is offset by the impossibility of givenness in the theological sense. Instead of treating the logia of Jesus with the hermeneutic suspicion that is explicit in philosophers, and carried out in secret among priests, who recapitulate it in the most insipid fashion, one would do better to treat them (and, with them, those many speeches of his that have not been authenticated by the Church and the official gospels) *as* axioms of a new kind, as oraxioms uttered under condition of the subject-science. On this basis one might hope for a renewal of the theories of interpretation and exegesis of the christological, or even Christian, corpus.

MESSIANIC ADDITION OR THE MEDIATE-WITHOUT-MEDIATION

Christ deserves the name of Mediate even though lacking all mediation—this is obviously a paradox. Jew and Greek could be combined only through the mediation of Christianity, and not prior to it. What was required was a law of mediation that Paul and the philosophers (Hegel) interpreted as "transcendental" in a broad sense, a law organizing the Jew and the Greek, but with the Greek mode dominant (despite the recognition of Jesus's Judaism)—whence the Jewish revolt against institutional

Christianity, which had ultimately made the Greek element dominant. Jew and Greek were "superposed" by a poorly understood form of Christian superposition. It was already compromised, the superposed elements being already reflected in the superposition itself, conceived as trinitarian. There is indeed a nonsubstantial and subjective thought-event at the origin of Christianity, but one that has been recuperated as the simple logical form of the Logos or the Word, still permeable to substantial religions. Christianity tried to redistribute the Judeo-Greek according to what it believed to be a new law of mediation, which in reality was only ever trinitarian. This ensured that the Christ-event would go on to give rise to prodigious philosophical developments.

But the force of the Jewish rebellion now reversed the hierarchy, and subjected the Same to the breach of an infinite transcendence, the extreme form of which we find, for example, in Levinas. How could this Jewish reversal of the Same by a transcendent messiah be partially recaptured as the Same or as the Mediate that we call "Christ," given that the transcendent One has been thus been foreclosed to the Greek? How could Christ be the transcendent One who has become Mediate? How could God become Christ? He is, on the one hand, without mediation, torn from the Greek. He is the Jewish messiah turned mediate, and therefore forever without mediation. Idempotence is the property that suspends his operation of mediation and recovers the Same. It is the operation of the Same, but contains the neutralizing suspension of its mediating nature. It is indeed the Same that conquers the Other, or the Other that can be "quantum-theoretically" added to it, that remains the Same, without taking stock of the mediation. The transcendence of the One or of the Jewish Messiah is recaptured in the law of the Same, in its immanence, with the Messiah becoming immanent or lived and ceasing to be divine, transcendent, or absolute Law—that is messianity.

The meaning of the Christ-event was probably falsified by Paul, or profoundly transformed so as to give rise to a tradition. The Mediate or Christ involves a mediation that could never be an auto- or even hetero-mediation; instead it falls into-immanence. To speak of the generic Christ, one must forge the concept of the *Idemmediate or Mediate-without-mediation*, like that of the *Idemscient*. The Mediate is the Same, neither identity nor alterity. For the first time, on the borders of philosophy, the Same finds a dignity that protects it from its traditional avatars, from Parmenides up to Nietzsche and the thinking of difference. For the first time, the Same does not return, but is deployed in itself, in its own immanence—this is messianity.

No transcendence can be correctly opposed to messianity, as a law of the world, because messianity strikes it down or depotentializes it. At once less formal and less empirically diasporic than the Jew, more immanent and less circular than the Greek, it is a semiformal legality woven into the lived that it includes and suspends, a materiel formalism. Neither singular nor a modality of the all, it sub-venes as an exteriority immanent to the world-subject, and it can be added to the entire ontico-ontological sphere while implying its transformation. The generic Christ is universal-without-totality, and for this reason inter-venes in the Greco-Jewish differend laterally, and can be added to it without negating it. Christ is the *addition or superposition of messianity to history*, which, through indiscernible interference, transforms but does not destroy. The unilateral way of Christ is, therefore, not exactly a formal method for resolving all problems, but an algebraic operation that includes them, so as to submit them to interference and render them indiscernible.

FROM THE LOGIA OF JESUS TO THE ORAXIOMS: THE MESSIANIC PROGRAM AND DOGMA

The logia of Jesus Christ, to which we could now add the apocryphal Gospels, have been ontologically "overinterpreted" by the philosophers who already inhabited the Greek Church Fathers and referred back to rustic, popular wisdom with disdain. A good framework for interpretation was not at hand; it was either sagely-Greek or popular-Jewish, and Christianity had invented rhetorical techniques to support both at once, even though Christ was not the result of this mélange. Even recently, in an infinitely less Greek fashion, Jesus's formulae on living creatures have been grasped as having to do with individual egos (M. Henry), though the only new sense they have is as generic, for the interfering and transindividual liveds that we are qua faithful in-the-last-instance. Let us change our hypothesis and treat those utterances as spoken rather than written, lived rather than logical—and yet transmissible—oraxioms, brief and fragmentary and thus indisputable and prior-to-first.

A science is obliged to build its foundations on a constant that defines a domain of objectivity, whether this be a logic and whether logical principles for philosophy itself should it wish to be a science, one or several natural properties for the various physics, an axiom-form or set-form as constant for mathematics, or a lived constant for a human science of religions founded in-Christ. We know that practically, for this science, we need (1) prior-to-firstly, a form that is not logico-Aristotelian but algebraic and determinative, idempotence, and the imaginary number; (2) a lived materiality (the materiel substance of messianity and of faith); (3) likewise a logical form, since it implicates some philosophy; (4) an axiomatic form since it implicates a rigorous science in its means. These various aspects are indivisibly interwoven in the statements of this science, which we have generally designated as lived axioms or oraxioms, and which are the faithful sayings, the "faithful axioms" of messianity. In-the-last-instance they express messianity while having recourse to the theologico-philosophical language that they transform. Each of these oraxioms is a generic quantum of expression or enunciation, not a conceptual or discursive entity, an atom in the transcendent sense, but a discrete and indivisible quantum of messianity. It is at once a drive, the raising of a cry, and the clamor of faith, the exclamation of a mystic. We shall not give examples, having treated more concretely cases drawn from mysticism, in *Mystique nonphilosophique à l'usage des contemporains* (*Nonphilosophical Mysticism and its Contemporary Usage*).[1]

We can now specify the essence, as messianic axioms in-the-last-instance, obtained by idempotent superposition, as has been noted, of the words of Christ and of theology. They are not logico-mathematical axioms, precisely delimited in a formal space; they are generically lived, having as their content the transcendent terms of Greco-Judaic theology, but transformed into the state of waves or interference, and therefore indiscernible. Into a wavelike state? One must not let oneself be taken in by the toofamiliar physical or material intuition of a wave. Wavelike here designates a schema or a form that is not necessarily a visible phenomenon of the sonorous or liquid flow of concepts, even though speech may be just such a phenomenon of flow and streaming, especially if one modifies or varies the speeds of physical flow in the manner known to mystics and ritualistic priests. The phenomenal mark of the wavelike as a schema that can sustain superposition is, above all, the interference of meanings or concepts.

It is a matter of projecting or making use of the conceptual kernels, and especially theoretical kernels since it is theology that is at stake, so as to bring them into superposition and interference, and of thereby creating new blocs of unlocalizable or indiscernible meaning. Interference is above all the interference of syntagms or of the corpuscules of signification that are statements. Now, the mystics (the Russians, in particular, rather than the Rhinish), the Hesychasts or Glorifiers of the Name, take the emission of thought and even of words to the limit of the discernible, for example, repeating the name "Jesus" or controlling the emission of breath. Speech obviously does not have the same power to synchronize and "spectralize" voices that we find in music. But perhaps it is from this general perspective on the interference of gestures and words, of gestures and speech contracted in time or even in space by the repetition of the same formulas, that one must understand an entire microritual dimension, and not only the macroritual of liturgies. On the condition that we distinguish at once between the mythological or macroscopic origin of rites and their properly microritual destination, which bears witness to an entirely other perspective. But the most fundamental thing, and what distinguishes our axioms from mathematical axioms, is that they are messianic *in-the-last-instance* or prior-to-priority. Our axioms are not only first from the point of view of their objective content; they are also prior-to-first from the lived-formal point of view. It is a question no longer of speaking conceptually of messianity in an atomist fashion, but of speaking in a "wavelike" fashion, with waves configuring concepts. Given their prior-to-first root or source, which is idempotence, they are "oraxioms," axioms generated as the oracles of this algebraic Pythia that is messianity. That science has an oracular and not only an axiomatico-deductive aspect is something that will be admitted by all those who practice axiomatic freedom extended to quantum, "wavelike," flowing idempotence, no longer lamenting an irrationalism that is meaningless here. An idemscient statement, at once lived and generic, the oraxiom is the indivisible duel, the generic quantum of unilateral formulation that is the property of no ego, but strictly speaking only of a "we" as quantum of expression or unilateral formulation that puts an end to a still atomistic axiomatic of mathematics or philosophy. Messianic immanence speaks through oraxioms that are distinct from the geometrico-transcendent axioms of Spinoza or of Judaic verses, from the conceptual atomism of the axiomatic (Fichte or the young Wittgenstein), and from the logical atomism of their avatars, and perhaps even, in part, metaphysical tautologies, from the first Greeks up to Heidegger. From this perspective, Jesus's *logia, which is to say, the collection of his words for the poor*, should be understood as formulae that are emptied of the sufficiency of the Logos, and that express an "impotence" of the Greco-Judaic, as duel (or binary-indivisible) quanta. They are sub-veniences, or, in quantum-theoretical terms, "Feynman histories" or "paths," philosophically incomprehensible—oracles in the memory of the humans to come—infinite and indiscernible words that traverse history and the world like an eternity that is no longer against the times.

If their formal aspect is wavelike, it is not just a matter of no longer speaking of the event or the meaning or the message in well-formed conceptual sequences, in an atomist and mechanist fashion. These axioms must be liberated from the Greco-Judaic, by their very emergence, which is not meta-Greek or meta-Judaic but rather "non"-Greek and "non"-Judaic, obtained through a quarter-turn or as an imaginary number. Quantum thought is a nonpositive act of thought—not only a set of discontinuous algebraic operations, but a real action rather than a structure of being, an immanence rather than an uncertain transcendence. In sum, we could say that superposition is superpositional-(of)-itself, which is to say that it is without-relation, without even a transcendental relation, to itself. It does indeed form a consistent though generic self, a Last Instance, and not a derived modality of consciousness or of the autopositional ego. The superposition of states yields a thickness that is incarnate rather than incorporated, a new, altogether immanent state, a materiel rather than a materialist spirituality.

The axiom as quantum of expression or unilateral formulation thus puts an end to the still atomistic axiomatic of mathematics, philosophy, and theology, with its axioms of the Trinity. From this perspective, the logoi or logia of Jesus Christ can always be taken as simple and popular figures of speech, ways of making a potentially complex message understood. There is therefore always a risk in rephilosophizing them at will, giving them a retrospectively explicable meaning in a theological fashion. But even more profoundly, theology is just a hermeneutic aid or a variable for oraxioms to be produced, these words being grasped as formulae that indicate in negative a *debased word*, emptied of its sufficiency, and that in fact express an ever-present Greco-Judaic "impotence." They are duel quanta, and in this sense should also be read as apocryphal formulations.

Finally, what does it mean to "philosophize in Christ"? To ordain Christian philosophical discourse to Christ-in-person, but on the condition that we make Christ the point where the Christian discourse to which he is foreclosed becomes irrelevant or impossible. In this way the Lutheran formula acquires its full force, and can be reworked into that of a "science-in-Christ." Christ is the source of life or the lived of the axioms, not a lived abstraction but an abstract lived, "formalized" (unilateralized) with respect to the Logos and the Torah. Christic science is thus a practice or rather a messianic "underpractice" of evangelical statements and their transformation into oraxioms, the constitution, via its very weakness, of an ultimatum-gospel that will have definitively ceased to presume to represent the world for the faithful or the "living." Cease treating the Scriptures, whatever they may be (Hebraic, Islamic, Christian, Philosophical), as sacred and fetishized texts. Invent your Gospels, those that the Churches ignore or that they are obliged to bury in the desert. The force of the Gospels is not "evangelical." Do not arbitrarily and viciously interpret the texts, but adequately transform them as a function of the christic constant, make them the vectors of a nonaction that is entirely one of messianity, rather than the means of a constricting proselytism. Fundamentalism in all of its forms, texts, dogmas, sects, and churches, with its parochialism, communitarianism, nationalism, and sectarian dissemination—here is the Enemy of Enemies, the Great Harasser, the Universal Inquisitor.

Eight

THE GENERIC SCIENCE OF THE WORLD

SUBSTANTIAL RELIGIONS AND FORMAL OR GENERIC RELIGION

The importance of Christianity and Judaism in the creation of sciences is a theory and a problem for historians, even if in reality their impact was only ever indirect. But we take up this problem in another way, in relation to the creation of a science of monotheisms and their theology. This genesis that we seek is obviously distinct from Plato's mathematico-philosophical creation, which takes place within it, or is an internal but particular realization of it. In any case, each attempt at a science is a rupture of myth, Plato with mathematics as science of the heavens, Galileo with physics as science of nature, Freud with psychoanalysis as science of the individual imaginary, Christ with the underdetermining placing under condition of those world-thoughts that are religions and what remains of them in their theological structure.

If Judaism is one of the conditions of the birth of modern science, as has been maintained, we hypothesize that this took place necessarily through the Christian mediation. The same goes for Greek paganism, which is a valid religion like any other. One invents philology, exegesis, and commentary, the other the axiomatic and deduction; these are frameworks of means for sciences to come, but do not at all provide the theory-frameworks or the science-continents in which they must be reinserted, possibly in relation to the "world as such" of which Christ (rather than Christianity) is the invention and the theory. We call substantial those religions that can, among other things, develop scientific knowledges founded on the unmediated primacy of the divine (even a polytheist divine), but that cannot directly furnish the model of a science of the world and thus cannot place it under condition. These religious images are globally "transcendental" structures—that is to say, structures with three terms or poles, historically diverse but sufficient for the type of invariant we seek. They are "substantial" or intraworldly religions that are on good terms with metaphysics, which they deform without suspending its transcendental kernel, as a function of their type of divine transcendence. Now, to put this transcendental structure of the world under condition, it is necessary to have invented a "form" or a formal paradigm articulated on a generic duality, man/world. This does not necessarily establish the technique of modern science, but promotes its generic character. "Formal" does not signify a formalism of the logical and mathematical type as usually understood, but a generic universality of the real, a human-oriented generic universality—what we call a materiel formalism. Christianity is the "formal" and "materiel" religion that invented, in the old religious context that served as its placenta, the mediate-without-mediation as unilateral duality and as messianic lived. As generic, it dedicates science to the world and to its theologico-philosophical structures rather than to the heavens or to signs, to human genericity rather than to the positivity of sciences-without-subject. It is only in Christ that we have the supra- or metatheological destination—nontheology is not a metatheology or a plan of salvation. What remains is to attach it to contemporary theoretical means.

THE ABSTRACT, JUDEO-CHRISTIAN AND GRECO-CHRISTIAN PROBLEMATICS

Let us eliminate straightaway the Judeo-Christian and Greco-Christian problematics. We are not interested in these overrestrictive historical problematics, in the broad sense of the relations between Judaism, philosophy, and Christianity. Judeo-Christianity is often a nostalgia for institutional Christians, a construction often judged to be phantasmatic, retroactive, linked no doubt to chance phenomena of theoretical and political exchanges between early communities, denied by many Jewish historians who refuse to be absorbed or sublated by Christianity. The time has not yet come to awaken what is a desire for origins or for a common root, a nostalgia for the Jewish emergence of Christianity that would historically and dogmatically complete it. History ceaselessly bifurcates at points of violence, rejections, and captures, and Judeo-Christianity is above all something that is at stake for the Christianity of the Church and seems destined to remain, on the Jewish side, a wish, certainly a very pious one. No ecumenism other than the political strategy of ulterior motives is possible between Church and Synagogue. Thus our point of view is not historical and continuous; it does not wish to and cannot enter into these so-called ecumenical debates. It is scientific: what can Judaism as material or variable contribute to the building of this science for which even Christianity is ultimately also a material or a variable? Let each remain faithful to the faith of his people and to his upbringing, Christ is another affair and another faith; fidelity is neither vernacular nor historical. The problem is not really any longer that of the Christian interpretation of Judaism, to be accomplished dialectically or otherwise, in a Pauline way or otherwise. One should not forget just how profoundly Levinas, for example, irrevocably closed the hiatus with the Greek and Christian grounds that he *exemplifies through the Hitler-event as philosophical event.* A perhaps excessive exemplification, but it is necessary in these matters to think with the excess that alone can instruct and prevent any theoretical reconciliation—*theory is not the life that contents itself with doxa and generality in the guise of genericity.* Attempts at synthesis, even very controlled ones (Derrida), if not incoherent, certainly can satisfy neither *philosophers* coming from a Greek background nor Jewish *intellectuals,* and still less the *faithful.* With Christ, a fracture of an entirely other order took place, and was announced by him; it is irreversible despite all efforts to suture it. We have admitted this fracture, though it is covered over with a transcendental appearance and perhaps more still with an immanental appearance.

Of course, this problematic must be extended in the other direction, beyond Judeo-Christianity, all the way to Greco-Christianity—which is equally imaginary if it is isolated and becomes a religious gnosis. But all of these mélanges are but materials to be transformed in order to understand the Christ-event theoretically. As impossible as this theoretical construction may be from the philosophical and theological point of view, here we undertake to construct, in the name of non-Christian science, a christo-fiction, real but philosophically and theologically unverifiable. There are in the above, and there will be in what follows, "christo-fictional" statements that will be received as theological fictions but that are nevertheless true, even if indemonstrable within the theological framework.

THE AMBIGUITY OF CHRISTIANITY

To understand Christ as "factum" of faith or as mediate-without-mediation, we have had to go back to the Greek and the Jew—in an external, substantial, and synthetic combination not as practiced by the Church, laying claim to its double "heritage," but as controlled by two quantum principles that are the refusal of all mélanges.

Substantial religions are autonomous and of a saturated consistency, they enjoy the independence of ancient roots. But Christianity is not a stock or a root—from this point of view it is more like a branch or a graft. Only the Jew and the Greek are our origins, not Christianity, which comes too late. The danger is that Christianity tends to establish itself as a root equal to others, to make itself a "religion" or even a "Europe," something that at the beginning it was not meant to be. Its strength is to be a form that prevails over substance, a mediation, strictly speaking, between Athens and Jerusalem. In relation to the Jew and the Greek, it is almost empty or without content—it is just a miniscule historical and sectarian event that does not change the givens but the very form of the game; it presents itself as a new law or logic introduced between conflictual terms. Apparently the Jew does not have the will to synthesis either, but is a pure form of repulsion, a refusal to be assimilated; but at the same time the Jewish form of proselytism is entirely a defense of its territory and its substance alike. Greek polytheism is substantial and, to that extent, oriental; it defines itself using terms with little form, or that play the role of a borrowed form, like the metaphysics of the One and the Multiple; it subsists with a hybrid or mixed will to proselytism, philosophy. As for Christianity, it diverts the work of Christ and easily becomes militant and missionary, its expansive force coming from this in-between situation and from the possibility of synthesis; it carries a certain power of universality that is not yet saturated, and in it the form of synthesis tends to produce its own religious substance to be imposed. To simplify and make a symptomatic link with our problematic, which is that of the idempotence or algebra of Christ, Christianity, which is historically an a posteriori synthesis, presents itself as an a priori synthesis, and in this way tries to encroach upon the other religions.

This religious takeover is not at all what Christ called for in his most unambiguous sayings, or at least he did not wish it in the authoritarian form that it later received, particularly with Paul. Christ is not the capture, disguised as "heritage," of the Greek and Jewish principles. If he transforms them, it is through an immanent procedure of the preemption of the lived of Jewish and Greek subjects, which he carries into quantum-oriented genericity. We seem to give priority to Christianity over the other religions, but whatever religion it has become, its essence is to be a force that wishes to be fundamental rather than a religious substance. This is why, if it is seen-in-Christ as we propose, it becomes a nonreligious force empty of all proselytism—messiahs are actual, even if one is tempted to say that they are "rare." The gnostic Christ is the prior-to-priority over all religion, which implies that he is freed from Christianity itself. We distinguish the christic stance from Christianity, which has only been able to borrow from existing religions so as to constitute this mixture that ceaselessly betrays Christ. It is obvious in its conditions that Christianity can only claim to be a science of other religions on the condition of renouncing, through generic faith or fidelity, its own beliefs.

THE GENERIC QUANTUM AND THE SCIENCE OF RELIGIONS

Generic quantum theory is not positive physics, but is said of the sciences that imply man as subject bound by the world. We must modify our concept of the quanta of action in order to give it a human orientation.

A scientific constant cannot be of the order of a subjectivity or a transcendental consciousness, nor of an "objective transcendental"—this is mere wordplay. The constant must first of all be scientific, like an objective law (Kant would say *quid facti?*, an a priori algebraic fact), and *to this extent* it is lived through preemption over the subject or philosophizable doxa. It is what we call the subject-science as "objective" nonreflexive lived. The idea of the generic is that every subject begins prior-to-priority through a scientific law. This is certainly an abstract or incomplete moment, and it must be fulfilled qua lived. Religions precede man, but man precedes prior-to-priority the science of religions—this also is the meaning of gnosis. The All is preceded prior-to-priority by science, but the All must follow and cede its aspect of the lived in order to complete the generic. The subject cannot support itself all alone through its own transcendental forces as constant. What maintains it constant and open is either a law of immanence, but a scientific or objective one, or the transcendence of the Other. To maintain this openness, one has the choice between the Judaic transcendence of the Other that affects man in exteriority and the immanence of a law as invariant or constant algebraic Same that "hooks up to" some nonegological lived. For our part we rely on the contingent rigor of science rather than the faultless rigor of God.

Michel Henry has formalized the minimal invariant for all possible philosophies of Greek and non-Christian, non christo-centric origin: "phenomenological distance" or exteriority as transcendence of phenomenal depth, a new triad. But this philosophical invariant is incapable of forming a generic constant of the scientific type; through a too-simple opposition, it makes possible an immanent ego as transcendental, which apparently has certain aspects of a scientific constant, but still a philosophical one. For phenomenological distance is an invariant principle, it is never annulled; the transcendental thus cannot risk being flattened onto the earth or the empirical. So what keeps the transcendental opening of Being open, preventing it from falling back on itself? What is it that maintains the transcendental dimension and makes it a quantum of phenomenal distance? Understood as transcendental, this constant is a pure supposition, for the circuit of being comes back not only on itself but into itself, from the exterior and the interior, or is flattened ecstatically in itself; it is in reality a circle that dissimulates itself. This quantum is really a quantum of autoimplication, a constant itself autoimplicated, and that must therefore remain in itself. It is not yet a scientific constant, which would have to be a stable but noncircular structure.

Christ invents the quantum of faith. He presents himself as generic (and thus unilateral) duality, as the necessary condition under which every life is led and reproduced. Faith, action, love, or power suppose such a conception of the quantum. These are obviously neither physical quantities nor philosophical differences: the generic quantum shows that the acting of man is "quantifiable" without passing by way of a hard physical quantification. Faith is not an "act" of faith but in itself a continual vectoriell flux that brings about the cognizance of messianity. There are quanta of faith, of action, or of love that are unilateral and associate two elements that are unilaterally heterogeneous, more than simply unequal or differential. The generic is duel (binary), but because it is quartial (neither singular nor plural, neither one nor multiple, but complex like the "imaginary number" or, more intuitively, the quarter-turn) in origin and is distinguished from the physical quantum and from philosophical difference alike. The Duality of the Father and Son reduced not to a simple transcendence and overpowering unity but to an immanence that suspends the Father, underdetermines him, or makes him fall into-immanence. The Trinity is obviously the return of the philosophical machine, which does not want to relinquish its property, and prefers to place the Father back into the transcendent(al) circle, with the Holy Spirit to bless this synthesis. This trinitarian triangulation with the holy familialism that follows from it flattens faith onto transcendent belief. The algebra of idempotence, on the contrary, is valid everywhere where the subject is involved alongside science as generic, and above all in philosophy and in religion. The amorous lived, for example, is duel without being analytic or synthetic, but just "neutralized" in its ego, giving rise to a science of lovers as generic subject-science.

Nine

INDISCERNIBLE MESSIANITY

AGAINST UNITARY REPRESENTATION

To extract the christic kernel from its outer layers, which have been validated as Christian? Certainly the Church has seen other heresies, heresies great and small—philosophy, too. Among contemporary philosophical heresies, it was Derridean deconstruction that tried to think in terms of dissemination or Judaic punctuality; the Deleuzian construction, on the contrary, in terms of oscillatory fluency. But both remain caught within the philosophical horizon, within the conceptual and semantic atomism of the presupposed-and-defeated All. They remained within this framework without giving themselves a new paradigm of thought. In each case there are torsions, deviations, leftings, and rightings (*gauchissements et droitisations*) of an orthodoxy of thought or a theological image of faith, an image that was put in place by forced marches, by a history entirely formed of discursive, physical, and mental violence through which the Church and the Academy constituted themselves, as a result piercing through this chaos of forces. But for us it is a matter of another project, of a science, an idemscient knowledge whose field is that of any religious object whatsoever, a field we have delimited by its extremes, the pagan Logos and the Judaic Torah. The kernel of the science-in-Christ is not any conceptual unity whatsoever, precisely not a conceptually atomic kernel. It is generic or duel, it cannot be said to be analytic or synthetic, or plural. It escapes these logics without having recourse to a negative theology since it transforms all of these predicates. Negative theology and philosophy have sometimes delivered Logos to self-hatred, rather than placing it under an idemscient procedure without negation or affirmation, under nonacting or nonbelief in some way. But whether affirmative or negative, they form mélanges, in the name of the All or the Absolute, of conceptual atomism and wavelike fusion, sometimes in real oscillatory machines (Deleuze). This mélange supposes the two styles to be separate and unitarily unified, whereas the quantum point of view also utilizes both of them, but without mixing or identifying them, rendering them indiscernible as superpositions. For example, we can no longer decide whether "Christ" marks an evental source-point in history and theory or instead outlines a transversal in philo-theological space. In reality, Christ is a vector of messianity that addresses itself to the world and that comes from humans into a "space" no longer defined by philosophical coordinates (ecstatic-horizontal transcendence and ecstatic-vertical transcendence, ontology and theology).

FROM THE CORPUSCULAR TO THE WAVELIKE: BELIEF AS THE DESTRUCTION OF THE MESSIAH

These retrospective theological and philosophical interpretations of Christ, his reduction to his "macroscopic" or worldly role as a founder of religion, are forms of the pure and simple destruction of the being of the superposition of Christ. All to a greater or lesser degree try to force the foreclosed-being of Christ and to theologize it. In quantum terms, messianity is not a flux that is numerically "one" and still corpuscular; it presents itself as a "unicity" but it is constituted through the superposition of liveds in a "wave packet," in the classic expression. So that all of these interpretations represent a veritable "reduction" of the packet of liveds that constitute the Christ-event. The Church, in particular, is the sum of retrospective dogmas and macroscopic constructions that destroy the christic superposition. The development of dogma is the systematic takeover of the new messianic and subtheological world. It is not the Church that is eschaton, it is the actual Christ, his active actuality contradicting neither his virtuality nor his ultimacy. There is an actuality of existence or of the effect of the eschaton, and the generic Messiah is its content. Radical immanence is obviously not the actuality of the present of the Idea, of Pathos, or of Life, which are becomings rather than under-goings, but the actuality of the messianic coming, which uproots the mediatic actuality of religion.

We shall thus avoid reducing Christ as Messiah to an event, to a punctuality or a cut, whether multiple or single. Messianity is an idempotent over-flow (*dé-bord*), (an immanent binary), a vectoriell wave or force phenomenon, but not the phenomenon of a point in the milieu of transcendence. The "Messenger" is not a specular image of divine transcendence. On this point, Judaism, too bound up in its spatial and temporal diaspora, has remained within a certain mythology of transcendence and has invented little. A point-One is empty by definition; it is, strictly speaking, a concentrated repulsive or retroactive force. A science of messianity is more interested in its materiality as wavelike or interferent lived, the plenitude of its stance, which excludes punctuality and atomist representation. For we who formulate axioms using philosophy (but only as symptom), there is indeed an operation of philosophical interpretation, but one that is neutralized in its aspect of ecstatic transcendence.

The One that we have previously used to speak this immanence that escapes every ontological norm would not alone have allowed us to understand immanence in itself, or only across many hesitations in writing and the simplicity of a causality through simple retreat or subtraction. Already we had to rework the One in new idempotent equations such as One-in-One or One + One = One. We had to make of idempotence the immanental neutralization of transcendence, not its determinate negation or indeed its immediate negation (Henry). Speaking apophatically of the non-One or of immanence in general, we did not yet have at our disposal the algebraic matrix of messianic immanence in its autonomy and the positivity of the Same obtained as superposition. This "algebraic" matrix of idempotence must now be incarnated in the lived, in order to yield the christic constant or quantum of faith. This is to distinguish the prior-to-priority of the Same and the priority of the One and/or of the Multiple whose philosophies thrive on debates. We thus no longer prioritize the characterization of the "nontheological" science through the will to do for the One what Heidegger did for Being—this parallel may correspond well to the stage of interference affecting transcendences, but not to a true theory of the immanent or generic or messianic wave.

We may therefore make a distinction between the two aspects of unilateral duality: on the one hand, the lived immanence of the generic Same (the Last Instance or the subject-science) and, on the other hand, the transcendence that, from being theological, has unilateralized or fallen-into-immanence, having lost its corpuscular and ecstatic sufficiency, standardized in the notion of God as "omni-." There is no doubt a transcendence in prior-to-priority of the generic stance; it is not denied absolutely, but is an ascending that is without ecstasis in its vectoriellity, and that only receives idempotence in the guise of a weak or generic transcendence.

Messianity is a type of event that is unique each time but is of the wave type or, strictly speaking, of the flash type, on the condition that we treat the philosophical flash as a transfinite flux of immanent light. It is symptomatic that the most radical immanence that the philosophy of Christ has imagined (in Michel Henry) is doubled into two equal aspects, an ego-point and an oscillatory flux of autoaffectivity that is closed up in this ego and that fills it—thus, between a conceptual atomism (the ego is semitranscendent or transcendental) and an oscillating matter, that of affective tonalities. Once more the ego or the center carries it on the messianic wave, whereas the property of idempotence makes of the immanence of messianity a vector, and not a circle, which it underdetermines; *the immanence of that which does nothing but come messianically must be sought in the greatest depth of the "without-return" or of the Resurrection of Christ.* The under-going of Christ is the only real possible content of his mythological "return," the only memory of his messianity. The messianic wave contains no atomist reference or intuition, is not conflated with the closed-up oscillatory, with the symptoms of divine transcendence. It is all of classical theological representation, and in particular trinitarian representation, that is inadequate. Fallen back on messianity, it would destroy its very principle of idempotence. Idempotence (idemscience . . .) on the contrary suspends the reciprocal mediation of God and of man. One can add the same term to it, or an Other, but since it brackets out the operation, it remains the idempotent Same. Consequently, the sacrifice of God is programmed in the form of neutralization, of his bracketed-out sufficiency: God himself is reduced as variable of messianity.

THE ACT OF NONACTING AS ABASEMENT OF TRANSCENDENCE

Idempotence is thus the capacity of an operation to produce the "same" effect, which confirms its identity, whatever the applications may be; but this formula is eminently ambiguous, and is the occasion to distinguish Judaism from Christ once more, to differentiate two types of acting or of messianity. To remain the same despite external operations of the variation (addition and subtraction) of predicates, properties, or cases of which it is the object is the capacity of an invariant or a transcendent One, the point that, through its transcendence, resists, despite external variations. This can be said of the Jewish people as an invariant that survives all historical pogroms and theoretical programs, in short "final solutions," but that at the same time condemns them to a waiting, a messianism that is empty and formal, as empirically concrete as it may sometimes be. But to remain the same also has an entirely other sense when it is a matter of Christ, who introduces a great disruption, or rather a Turn, which is that of generic messianity. It is no longer that which remains, or resists variations, and which is designated by these eternal historical variations as the Same, stubborn in its transcendence; it is the Same that resists because it remains "in-itself-same" through its superposition with itself, and that at the same time depotentializes adverse forces. Anti-Semitic variations serve to make that which resists appear, as an operation of glorious and stubborn transcendence, whereas messianity does not have to constitute itself *in history* as that which resists: it is from the start indifferent to whatever variations or operations may be added to it as its reprises. The Messiah does not resist aggressively, and yet Christ is our "hard-and-fast" fortress, a photon rather than a spark, reduced to vectoriell immanence.

Christ has sovereignty over the Weak or the Simple, and a certain nonacting in action. Nonacting meaning that he does not have to act directly—that is to say, to re-act re-flexively—not that he does not act at all. He does not have to act mechanically or metaphysically, making use of double transcendence in order to resist it. The power of remaining the Same is the resumed messianity that depotentializes violence; it is not the power to bear the countertime of history and to suffer its avatars. All the more reason to include or exclude them, to make of them a philosophical machine, a synthesis in a great theologico-globalized system, or else a triumphant machine of vagaries and chance, including them by disjunction or by negation in an affirmative diaspora, like an eternal return of Judaism. Messianity is neither affirmative nor negative. Even its resistance is not direct or immediately addressed against exterior aggression, but is indirect and operates through suspense. The ultimate christic act, the act of ultimatum, passes by way of a nonacting that is neither a void nor a waiting. The incessance of salvation means, among other things, that theology has not elucidated, any more than philosophy, the essence of messianic causality.

At its root, messianity is thus no longer a circle that constitutes itself by reconstituting itself, a production that includes its reproduction, as philosophical circles do; it is a resumption foreclosed-to-variations and repetitions, which are, from the start, neutralized or vectorielly added, an immanent flux that is equal to itself no matter what the variations may be, and that in a certain way flows "into" itself. Nevertheless, when it is completed as generic constant through the insertion of the occasional transcendence of the world, it can be said that it constitutes "itself" as indifferentiation, neutralization through interference, affecting double transcendence or traversing it via a tunneling-effect. But even in this case where messianity is in some way effectuated, it is still distinguished from the Same of philosophy, which autoconstitutes itself as Eternal Return or overpowering.

Messianity is constituted not as power of return or of repetition but as duality of repetition suspended or resumed. It is one time each time *one sole* unilateral parenthesis, which now affects transcendence itself as transfinite. Among philosophers two parentheses are needed around the world; for Jews and Christians there is one, through the transcendent subtraction of the other; but Christ needs none, and thus he can always add one depending on the occasion, but if so it remains immanent. Messianity does not put itself, operationally, between parentheses, through a transcendent act of constitution. But as infinite as it may be, once it has to do with the world, it is completed by one sole parenthesis that opens it unifacially, as Stranger, onto the world.

Unlike the concept that determines in idealist manner the experience of the materialist void, which subtracts itself from it and conserves it specularly, messianity underdetermines events by putting the world to which they belong under condition of transformation. We are neighbors of idealists only in the same way as we are neighbors of materialist or

Far Eastern philosophers of the void. The transformation here is one of ontological form, not one of material content. Doubtless under-going or underdetermination is an operation that does not directly affect reality in-itself, the Messiah has no goals internal to transcendence, but only an *immanent work upon* it: he indirectly depotentializes or sterilizes the dominations and the doublets that make the world. In other words, the act of an idempotent nonacting does indeed have a content that is a neutralized materiality, one that has lost its macroscopic autopositionality through complex superpositions; it is not a material void or an empty set. In particular we must distinguish between the empty set and the idempotent wave that *neutralizes* or underdetermines every positional content. This is the condition of the transformation of the occasion, which loses its sufficiency and becomes indiscernible or superposable. There is obviously a stratification of operations of this idempotent neutralization as a function of the structure of the object "philosophy," but it is an occasional phenomenon distinct from superposition. Idempotent sterility transforms the structures of Christianity and its theology, whereas the absolute void of mathematical set theory and the void of the dialectic, not to mention the atomist void, all conserve them without transforming them, just purifying them ideally, suppressing them, and reproducing them specularly. It is quite understandable that according to this problematic, external to Christ, the only significant event in the sphere of Christianity would be the intervention of Saint Paul.

UNTRACEABLE MESSIANITY

The event that we are concerned with bears the singular name of "Christ." But by definition of the idempotent fluxes that guide or mobilize innumerable particles of kerygma, of transcendence of religious extraction, messianity is a wave packet of liveds, or a complex of messianic vectors. Like every wavelike phenomenon, messianity is a set of pathways, innumerable and undecidable ways, possible pathways that are hazardous to pursue, without any identified traceability, or that leave traces only after the fact, once they have been detected or received in the so-called conscious element or the element of belief. Monotheistic religions, including the Christian religion, are encumbered with local Christs or imaginary prophets who are like discernable macroscopic representations of those messiahs who are the ultimate and indiscernible faithful—but become warlike rivals. The immanent packet of fluxes is also one and multiple, but belongs to the microscopic order, with neither analysis nor synthesis, without passing by way of a "Platonic" interlacing of forms. It is not one and multiple in the macroscopic order of metaphysics and its entities: the generic disregards genera in favor solely of man. From macroscopic anarchy to the unknowing of superposed messianic waves—gnostic knowledge that does not know—there are doubtless effects of resemblance between the orders. But we do not know the possible pathways of the messianic vector and of the kerygma (its "Feynman paths"), and this unknowing is a well-determined ignorance, which does not result from mélange and confusion.

Messianity has no mask (Nietzsche) and, inversely, no interiority (Kierkegaard), but only a vectoriell immanence. It is also not identifiable by the Churches except in the "eyes of the world." Neither time nor place can be parameters of it. There is no traceability of the messiah—this is a great achievement of Judaism that Catholicism, above all, has forgotten through paganism. But Judaism has multiplied and rendered apocryphal messianic advents, the better to exacerbate waiting and make it passionate. When Levinas philosophizes the trace, and Derrida follows him, it is a matter of what remains of it as memory and as past eternally present in the self. They do not venture into the most solitary desert of the soul, a desert of unknowing that accompanies the knowledge that we are, and whose cognizance does not suppose a prior trace via a remainder of Platonism. All that remains is the occasion of "theology" rather than the after-the-fact trace left by a trauma. Just as there are philosophers who refuse to break with opinion and common sense, there are religious confessions that refuse to break with pagan doxa, that are still nostalgic for polytheism in the form of "polychristic" representations.

The philosophico-religious problematic can be recognized by the way in which it functions on the invariant of phenomenological distance in alternating contrary ways, of more or less, of saturation or desaturation, whereas the least positive science that we require as a model functions on idempotence and noncommutativity. The immanental phenomenon of the generic lived is found at the end neither of an annihilating (and "desaturating") destructive distance—destructive, more exactly, of the idempotent "packet" of liveds—nor of one saturated with an overplenitude of sense or alterity, as if the distance were that of an alterity or of a transcendence, of an Other getting a foothold directly on either the lived or the claimant (the "interloquant"). The lived phenomenon is simplified and transformed, the simple in this case being superposition rather than saturation. The ego is reduced generically rather than interpellated, and the problem of intersubjectivity is no longer posed for it since, without being many or solitary, it is duel, the identity of a binary. The couplet Ego/Other is traversed by the Same and redistributed according to a unilateral duality. The subject cannot be interlocuted or interpellated, because there is no interlocutor, but only an Other fallen-into-Same but despite that always Other, albeit a weakened Other.

The faithful of-last-instance perhaps do not come forth like knights of faith but, for sure, like probable, indiscernible, or quantum thieves. Their probability is a probability in principle, because the religions that cultivate mystery so well have perhaps lost the secret of the secret—namely, that the most real of faith has been revealed to them as underexistence, as the under-going of an ontologically wavelike rather than substantial nature. If the messianity that is declared in the sayings of Christ is established on the ruins of theological sufficiency, it falls to us as the faithful to make it known—that is, to make of it a scandal, a paradox, and a folly.

Ten

SCIENTIFIC DISCOVERY AND REVELATION

GRACE AND HUMANS, A UNILATERAL COMPLEMENTARITY

The "foundation" of the Christ-science cannot be the same foundation that religions, in all their transcendence, imagine for faith: illumination, revelation, vision, or "inspiration," wind, breath, or light. Everything that is of the order of light, of the quasi-instantaneous flash, or of the spark is indeed a theoretical argument, but a corpuscular one that blinds itself: the theoretical as self-blinding. When light is made streaming and flowing, as with the mystics, the problem better reveals its true nature, its "wavelike" phenomenal content, even if this is often covered over, flattened, by the officialization of beliefs, images, and dogmas. The metaphor of revelation is a contraction of sensible phenomena that must be made explicit and rendered intelligible by means of an adequate science. Religious-type faith is a belief-with-closed-eyes, bad blindness rather than bad faith, and continues to project afterimages in transcendence—the shadows that people the celestial cave.

A fortiori, when it comes to making a claim to a science, and to thinking invention rather than revelation, what is needed—what has always been needed—is some scientific *model* or other supposed given as more rigorous than others, or at least as implying a cut with anterior or ordinary experience. We find symptoms of this in Christ, who makes use of a certain formal rationality (the Judaic) in order to surpass it, just as he uses an element of Greek rationality, mediation. But the science capable of explaining the wavelike contracted aspect of revelation and, what is more, the statuified images and the dogmas that accompany it as the corpuscular material of religions were not yet within his reach. Once it becomes a means or an integral part of generic faith, and once the old rationalities have been brought down to the function of mere models or productive forces, the introduction of science into the material of beliefs creates a revelation-effect not entirely unlike that which we find in invention or discovery, but which is of a new type. This introduction is obviously not without its psychological and historical conditions, but they are solely of an occasional order and do not constitute the process of invention itself. The revelatory aspect, now under immanent scientific condition, is reduced in its most transcendent aspects to the preemption of the lived by idempotence, recalling divine rapture or mystical ecstasy. It tears the subject from its individuality and incorporates it into the generic stance, a mystical body of which humanity is capable through its own means or those it appropriates. This algebraic rapture of belief transforms the latter by way of a new rationality which is that of messianity, that of the strong analytic and weak synthetic, without this involving any medium whatsoever, any equilibrium operating through exchange and compensation. Strictly speaking, we must not mix these two aspects, as if they tended toward each other; they retain the specific autonomy of their natures and, if they function together, they do so only within the apparatus of salvation that is Christ. The "opposites" or the occurrences of the idempotent term—here, Christ—are superposed, and form a complementary duality or a messianic machine with a broken or complementary circuit. If there is a return of Christ, it is thus an immanent reprise by faith of messianity, rather than the repetition of a pretraced circuit or trajectory.

What is the subject, if there is one, that brings about this resumption? Revelation as preemption of the subject by idempotence is an immanent or generic formalization of the worldly subject. It is not caught up in a circuit of action/reaction. The idempotent nonact finds some lived offered up by the transcendence that it neutralizes or "formalizes," which it thus manages to tear away from the world. Revelation is this tearing, which is consummated on the Cross, the Cross from which transcendence remains hanging like an old man's rags. All of this becomes generic when the material is that of the philosophical mélange or religious doxa that sets itself about questioning, intervening with, and harassing humans. Then the individual subject has no power to choose or to will qua generic. At best, he wills this or that condition, and cooperates as an occasion. But he is gripped "automatically" by the weak force, by nonacting, the generic indifference of idempotence, and begins to flow "in" it as "in"-Christ, the only grace now conceivable. It is enough that the world manifests itself in its manner, and that there is thus some lived "available" to resume idempotence as superposition or the superposition that is in idempotence. Humans as inhabitants of the world cooperate in this immanent grace, awakening or activating it; but they are a reservoir of passive acts in-the-last-instance for its operation. Grace, also, is a unilateral complementarity, and exits from the antinomies of its unitary concept.

The result is the foreclosure of messianity, or its noncommutativity with the philo-theo-logical subject, which cannot return. Precisely the "return of Christ" is impossible, it is one of the shadows of the celestial cave; noncommutativity prohibits it and can only declare imaginary albeit objective the psychology of mystics and of believers insofar as it is authenticated by theologians.

REVELATION AND DISCOVERY: SUPERPOSITION AND SATURATION

Positive-scientific discoveries such as the laborious discovery of a new constant, which sometimes end in a jubilant illumination, have some characteristics in common with revelation as a supposedly divine operation within the subject. There is even an experience of grace involved in the greatest of such works. As to the science-in-Christ or the vision-in-One, how is it to be situated between these extremes? It is all the more important not to conflate these three ecstases—it is not the same light nor the same enthusiasm nor the same localization, at least if the revelation called "messianity" is understood through an immanence that implies a minimum of unilateral duality between the instances in play.

There is, however, a paradox that allows us to distinguish immanent scientific discovery from the philosophico-theological "revelation" of a marvelous and illuminating exteriority, two things that are very often confused for each other. It is partly linked to the vision-in-One, which, as we recall, has to do with the removal of principles of sufficiency. How can a discovery be made by means of immanent mechanisms, as is the case for the science-in-Christ? It requires a new classification or sequencing of transcendence in its various heterogeneous moments. It is a question of breaking with its supposedly double character, its doublet symmetry, its oversaturation, of distinguishing between a simple or vectoriell ascender without transcendent object, a nonecstatic and autonomous ascendance, which remains under what we shall call (without explaining here) a certain waterline that distinguishes simple vectoriell pulsion from transcendence or its apparent existence as double. This ascender as pure vector affects doubled transcendence, drawing it into a simplicity stripped bare or "discovered" "in its naked functioning." What is discovered if not the simplicity of exteriority, the immanent or vectoriell essence of ecstasy become instasy? It is less an achievement or a fulfilling, or even a saturation, than a liberation—more an impoverishment than an enrichment. Scientific discovery is precisely not a revelation that would be re-covered by a second transcendence; strictly speaking, it is a recovering of the simple or nonecstatic vectoriell essence of transcendence.

Idempotent distance is thus distinguished from "phenomenological" or ecstatic "distance," from the depths deployed at the behest of the subject; it is contracted into itself and full of interferences but not saturated, completed but not closed through a return into the circle of the same. Unilateral duality allows us no longer to conflate an immanent "saturation" of faith via idempotence, borne by the fluxes of messianity, with a divine saturation via double transcendence that would close off messianity and would make faith into a fortress. Thus, revelation understood as christic is not entirely foreign to generic theoretical discovery, to the vision-in-One, which, as we maintain elsewhere, is a characteristic of the lived of knowledge. When the lived is forced to rank alongside idempotence, it is grasped, surprised by its discovery of the immanent extent of the almost infinite phenomenal possibilities delivered by dis-covered religion.

The revelation of the "in-Itself-same" "in" the Same, which is therefore not a saturated given, explains the feeling of oceanic flux and of an interiority of grace according to which the lived flows, runs, or carries one along. But equally, it explains the affect of exteriority or the "shock" undergone by a transcendence that is stripped bare and that is thus no longer that of a force hitting an object, but of a fallen into-immanence, like an internal collapse. The messiah that "I" am is the generic feeling of flowing across one transcendence to another, across dogmas and beliefs, without stopping at any as "last," since the generic is the movement that goes from man to man "under" the world or under the waterline, as if through a tunneling-effect. Such a superposition of discovery and of revelation in true invention is the secret of gnostic knowledge. It is the doing of generic automats that are neither singular-egological nor universal, but that we have called "quartial" rather than "partial." The Messiah, still less messianity, is not a true Idea to be imitated, nor any other datum or factum of consciousness; it is a messianic wave in which float liveds and dogmas in the state of "particles" or "clones." This continuum of the lived and of its multiple phases, certain thinkers or discoverers have managed to make it or have let it burst forth more than others. Messianity is the miracle of a logical (algebraic) form made entirely of the lived, a way to follow faithfully, a flowing objectivity such as Husserl, for example, sought for so long. Let us cease to await the Messiah a first or second time, and to imitate Christ. . . . Human intelligence does not imitate the divine understanding either in creation or in production; it is the ultimation and universion of these relations to the world.

CONVERSION: ULTIMATION: UNIVERSION

For the Christ-science, conversion is a religious appearance—objective no doubt, but meaningful only from inside belief; it cannot engender the most faithful of faiths. Conversion in the classic sense is the law of commutativity that is valid for the macroreligious world: a reciprocal conversion of God and Man as experienced by the mystics in the form of a twofold ecstasy. To the conversion of God to Man in Christ, a second creation, responds that of Man into God. This circle is a kind of theological eternal return of the same, and occurs in the Christian context founded on the couplet sacrifice/conversion. But if the sacrifice is that of God, as we shall hypothesize, then the messianic man who is born of the sacrifice of God does not need to be converted; religion, however, must be uni-verted outside of itself. As decisive as the idempotent addition of the two knowledges (Logos and Torah) is, the second algebraic principle, that of noncommutativity, establishes this addition in the space of Prior-to-priority or Last Instance.

The Last Instance is posited via an operation of so-called ultimation. Three claimants may contend for this function, and we have retained all three qua symptoms, a little like the "three sources" of Marxism: quantum theory and its microphenomenology, Marxism and its last-instance, and gnosis and its human knowledge of the world. Who "decides," then, to make the addition of two knowledges in view of the projected science if not, psychologically, a certain "subject" who is hesitating between philosophy and science, unsatisfied with epistemology, and who, exceeded by his own contingent hesitations and variations, ends up making the wager of a rigorous gnosis? This, in any case, relates to the psychological conditions of the christ-Science, and is not determining here. Moreover a science of theology contains, from the point of view of the latter, a more interesting movement of apparent conversion, which goes from philosophical or theological sufficiency to the scientific or generic stance. At this point, if neither science nor philosophy is problematic, being nothing other than what they are or claim to be, their combination, on the other hand, is problematic to the highest degree. But in this "turn" that would be the conversion to the generic, quantum theory imposes its principles from the outset, and takes hold of the conversion, which it transforms into a superposition. Superposition has an effect of uni-version—it tips philosophy out of sufficiency or makes it fall into-human-immanence. The subject has hardly operated this conversion before it is in effect already "turned" or uni-verted into a generic subject as an outcome of this addition. We say uni-version, but it is no longer a question of a "religious" and believing conversion to gnosis. If we suppose that a philosophico-religious subject decides to attempt this wager, it is itself immediately transformed, and cannot recognize itself exactly in the new man who emerges as the correlate of its operation. On the one hand it is displaced into-prior-to-priority, humanized—that is to say, generically formalized; and on the other hand it is deindividualized without losing all lived, losing just its egological sufficiency.

THE DETECTION OF THE MESSIANIC VECTOR: FROM WORLD-CIRCLE TO WORLD-SCREEN

Another aspect of this is the empirical, continuous, and slow invention of this science within the time of history, on the basis of the givens of theology and under its authority. But the wager of ultimation as superposition takes place in one single throw of the dice, where the subject is conserved only within its being-transformed as generic or messianic lived. Ultimation thus has two aspects: a contingent aspect as subject, and a necessary aspect, not in the sense of a sublation interiorizing contingency in the form of the fall of the dice, but in the sense of a superposition of the two religious Laws that produce a constant (the Same) including the contingent variable—in sum, generic man. The philosophical throw of dice as transcendental throw is replaced by the immanental throw or vector of messianity. It is the generic matrix as prior-to-first superposition that throws the dice of science and philosophy in such a way that they are added "quantically" (not arithmetically)—an immanental gesture, whose outcome is not a falling of the dice onto a table, but a throwing of messianic vectors in an imaginary quasi-Hilbertian space. Heraclitus, Nietzsche, Mallarmé, Deleuze, and many other lovers of philosophical immanence and of dice throws trapped within the circle of the world have disregarded messianic vectoriellity, for fundamentally Greco-religious or pagan reasons. The Nietzschean gaming table is a sort of transcendent dark precursor of immanence, limited and entirely belonging to the transcendental gesture. In the generic matrix, our two dice, science and theology, do not fall upon that plane of immanence that would obviously be the world. They form a messianic vector that crosses the cosmic wall, grabs hold of transcendences, makes them interfere, and thus creates innumerable and indiscernible fluxes or lived guiding functions of these transcendences. Never totalized or closed by the world-all, by the world-point, they are added only on a world-screen that simply detects them without their falling locally upon it and globally within it.

To capture through faith the hazardous effect of generic messianity, the world must abandon its transcendence, and must itself also become immanent like a detector screen. The vector of messianity underdetermines the world, or at least its philosophico-theological form, as a detector screen that belongs entirely to the "Christ" apparatus. The apparatus of the measurement of messianity is not external to it; it also is deduced under occasion of the messianic wave function, and within the limits of the latter's immanence. Generic immanence operates a nonvicious genesis through the transformation of this transcendence. It is an immanental deduction of the world-screen, not a transcendental deduction in the manner of philosophers—a thought without materialist topic, a radical base, a Last Instance without being a backworld. Evidently, the dice throw remains pagan; it is a conversion of dice to earth and heaven, to their reversibility as world; it affirms the primacy of the fall of the dice over their throwing, and forgets the prior-to-priority of the throwing as guiding vector of the dice that it makes interfere. Since the messianic Simple or the vector is not folded, but remains the unfolded that it under-goes, it has no need of a plane of immanence to prevent it from falling back into the void, but itself constitutes the radical basis, with neither plan(e) nor foundation. The generic messiah transforms the world-circle, the world-prison, into a world-screen that detects him by liberating him from the imaginary localizations and globalizations of religions, by transforming him as far as possible.

THE SUBVERSION OF THE MAN OF SCIENCE AS HUMAN TYPE: IDEMSCIENTS OR THE SIMPLE ONES

Philosophy is founded on the principles of identity and reason; what Christ contributes is of an entirely other nature—the principle of idempotence, or again, of simple or idemscient science; and a function or factor of the imaginary. Following the mystics, philosophers have also sought the "simple," but the simple cannot be an anonymous, transcendent, and punctual entity, a point. The critique of ecstatic representation must be followed by a critique of the punctuality of egological immanence—the nonecstatic must be of a wavelike nature. As for the gnostics, they know that humans are the Simple Ones, and define them by way of idemscient knowledge, repetition suspended or neutralized by way of reprise. The Simple is not a glorious or stubborn invariant, but the flux of equality that flows into itself as through a tunnel, without going beyond itself, a factum for those who have a knowledge of which they are not, at the outset, the knowers. The Christ-factum is an Idemscient brother of the gnostics, a subject-knowledge as index and function of its own announcement as variable. Religious gnosis had a subject, a messianic one even, but in a religious context. The oracular/popular style of some christic gnosis or of that under-logos must be brought to its oraxiomatic form.

We distinguish the Jews of scholarship (rather than knowledge), the Greeks of thought (philosophers), and the simple ones of knowledge who are the Idemscients or gnostics. The Idemscients have a simple knowledge, in the sense that it cannot reflexively know itself. Paradoxically, they are, as irreflexive knowledge, the human type closest to the scientists of the future, the creators and manipulators of automata. The intuition of science as in-One, the affinity of science and generic immanence in-the-last-instance, has always been the quantum-oriented inspiration of nonphilosophy. Its initial "intuition" not being philosophical, it is now developed with the contribution of quantum theory and in the direction of a complex and generic affinity with philosophy. Perhaps we ought to abandon the term "scientific" as too positive and technical, troublesome and undignified, a being of mass control, and along with it that of the "scientist"; perhaps they are too archaic and ill suited to the contemporary type of the man of science and should be replaced by the more exact term "Idemscient." "Idemscient" is the closest term to "man-(of)-science," "man for science (*homme à science*)," *homo sive scientia.* The gnostic man of knowledge admits almost the reverse, a science-(of)-man. Man is defined not as individual subject, but as the function of a superposition of science and philosophy.

The generic bears with it the miracle of this encounter of a paradigm that will become scientific and a lived that opens the way to humans harassed by the world. It refuses the humanist notions that belong to a transcendent, Greek, hierarchical categorization, with its famous distinction between *vita contemplativa* and *vita practica,* and whose ultimate fulfillment is the Nietzschean typology of the overman. The Simple unify man and science according to a new relation such that science, for philosophy, contributes a lived-without-life and forms with it a function of generic humanity. The Simple Ones are not scientists, they seek neither glory nor gain from their knowledge; and yet, if there are "scientists" in the sense of idemscients, these indeed are they.

Eleven

THE SCIENCE OF THE CROSS

THE RESURRECTED CHRIST AND HIS CHRISTIAN INVERSION

Many mysteries are condensed in the Cross, but some of them—those concerning its representation—are easily dissipated. The first is the mystery of its matrixial ontological structure. Its symbolic status also owes something to the way it is used and to its technical construction, in the crossing of the wooden pieces. Now, philosophers and theologians are known for thinking in two very different ways: on the one hand *ecstatically or pictorially* according to representation. But also, according to the objection made to them by certain mystics or, for example, by Nietzsche, for *obliquely* turning their gaze. Ecstatic and oblique—isn't this contradictory? Not at all: the ecstatic is realized through a division of directions and a torsion of the gaze onto oneself. It begins with an obliquity, and then the object comes back to it from the world, whereas it believes it has ecstatically reached its heart. And so they proceed to a transcendence they believe to be absolute but complex or falsely simple. Most philosophers have a spontaneous epistemology—they believe they are looking the thing in the face or in a face-to-face, even if this means attributing this straightforwardness to an ethics (Levinas); they say that they think the object itself or the phenomenon, when they always reach it definitively and without recourse to lateral adumbrations. It is appearance that is the vehicle of the oversimple notion of "phenomenological distance" that flattens onto itself its own operatory content, and that psychoanalysis simply ended up complicating. But psychoanalysis, precisely, contents itself with dividing the gaze, and thus still supposes the unity of perception that would remake or reconstitute itself in an imaginary subject. Just a little analysis shows that philosophy and thus also theology grasp their entities in the simplified form of an ecstatic transcendence, which secretly harbors specular dualities or doublets. Here we find a modelization through *appearances* that is the vehicle of an idealized perception fabricated according to Greek presuppositions. This is also the case when it is a question of thinking God or representing him. Theology, both ecstatic and oblique, is carried spontaneously to God as to its object par excellence, only then to think Christ in terms of God, never really doubting the ruses of its operation. It is not necessary for theology explicitly to evoke the Church Fathers in order to "think Greek": the simplest representation of God and Christ is already held fast in the interlacings of objectivation. Far from being simple or being complicated only theologically, the theo-christo-logical doublet is already ontologically complicated, and flattens onto itself the operation that it shelters. Feuerbach's Hegelian critique of alienation, restoring the autonomy of objectivation and distinguishing the finite individual in his relation to his generic infinite essence and in his relation to religious illusion, also confirms, like psychoanalysis, that the simplicity of ecstatic transcendence is only an appearance. But these two critiques, like others, remaining within philosophy as universal presupposition, content themselves with a decision in favor of the supposed unity of transcendence. It is inevitable in these epistemological conditions that still privilege (albeit from afar) the model of perception that the Cross, which is in fact the doublet of the crossing of perception, should become a symbol freighted with folly and philosophical paradox.

The second mystery has to do with its symbolic status. It is most often understood as a conjunction of two essences or two historical natures, of two beliefs, Greek and Judaic. This conjunction makes of the Christian either a "converted Jew" or a "converted Greek," either of whom must, for their part, reconcile or render compossible the two authorities. Do not this differend and the compromise solutions to it form the greater part of twentieth-century philosophy? We have the choice of converting ourselves to the other side, either to Judaism or to ontology—such is the misery of this hybridization of thought that remains in a state of division and hesitation that exports its own war.

The folly of the Cross lies at the heart of Christianity; it is the interpretation of Paul, a converted Jew, and one who understands it as the paradox of two irreconcilable languages that seem to call for a dialectic. The folly is obviously not a mere Greek paradox or a Jewish joke—to give his son for others, for the world, is completely absurd, worse than suicide—the folly begins here. The mélange of Law and Logos under the authority of one or the other is assumed finally by the Logos, and will never cease to resound within it and to trouble it. The folly of the Cross is also the folly of Paul's conversion.

As for Christ, he does not need to be converted, any more than he needs to be circumcised—we therefore cannot be sure that his real mystery can be reduced to the symbolic or idiomatic complexity of the Cross. It is the whole theo-christo-logical doublet, along with its architecture, that becomes problematic once we consider the "position" of Christ, if such a position is discernable, within the "artificial" space and time of an experiment rather than that of the created world and of history. Christ is perhaps not content with "fulfilling" the Law, but proceeds otherwise and frees himself from the old onto-theological framework. Something like a "death of God" more profound than his "moral" and "modern" death becomes possible. Should we not say, in a gnostic spirit, that Christ has come to "fulfill" God himself, to subtract him from his Greco-Judaic framework? That the Resurrection is the true meaning of the death of God and that the Crucifixion, that morbid obsession of the Church, has been overexploited in favor of God and his authority, and for the benefit of the Church?

On the other hand, if we maintain that the way of messianity, the only real way, has been inverted and theatricalized by historical Christianity, then we lay ourselves open to the charge of giving another "ontological" and "symbolic" interpretation of the Cross in terms of the Resurrection. If "conversion" as act of becoming-Christian is the determinist and worldly inversion of the Resurrection, a subjective and generic interpretation of the theo-christo-logical doublet is our objective.

Christ is the condition under which the believer can be uni-verted to faith, torn away from his sufficiency as Jew or indeed as Greek. Paul is cast down as Jew, illuminated as Greek—such is the folly of a collision that remains transcendent, like that of two worlds. He privileges the two sides alternately, but posits them together within a simultaneity for which the Logos is necessary, and which constitutes a matrixial duality. This is therefore already the multiplication, perhaps even the reciprocal multiplication, of languages, as a translation of two idioms into each other. But the dialectic remains solely an affair of a mélange that adjusts itself between languages that are contradictory at worst, opposed at best, and that does not bring into play the "real," but only logic, as Kierkegaard would say. The Cross lived in this way represents a transcendent logical or dialectical interpretation of what really happens in it qua generic matrix: doubtless already a multiplication of variables, but one that remains reciprocally commutative, because it remains within a philosophical horizon and does not become generic or follow any scientific model. In this form, the Cross represents the incomplete beginnings, still given over to religious transcendence, of the generic matrix. It is the ultimate triumph of God over Christ, of Father over Son, and not yet the becoming human or immanent of the Cross. Paul remains Jewish through the Greek and Greek through the Jewish. Having lacked quantification, he will lend himself, at best, to a deconstruction. Instead, uni-version or ultimation would be what we call the superposition of a being-overwhelmed or blinded and of a being-illuminated, far from being that weak collision of dialectic that will resolve itself, as usual, through a flux of new "didactic" writings and a planetary evangelization.

FROM CROSS TO MATRIX

The "folly of the Cross" is a mystery in a state of survival. Coming to us as absolute or transcendent paradox, it can be reduced to the radical or generic form, a unilateral as much as a human form of paradox, removed from the shadows that, it must be said, are metaphysical in several senses. What the Cross recounts to us is obviously first of all a legal rite with protocols and preparation, interrogation, judgment, and procedures for execution and the dispersal of spectators. And then it is a drama, a symbolic operation endowed with a theological meaning and a function in the plan of salvation. The problem will lie in not concluding the meaning from the image, following our determinist penchant. Usually the Cross is immediately interpreted symbolically in terms of the meaning that it received later; now, it is an object whose perceptual and intuitive perception is ambiguous and related to quantum procedures. A pictorial or symbolic image—never mind its theological or artistic usage, we shall bracket it out provisionally. A phenomenology of the Cross must consider it more profoundly in its *matrixial materiality*. It must be read as a dramatic story and a deep well of mystery, but we take it as a perceived phenomenon open to a physical or quantum interpretation. We mobilize categories other than the existentiell categories—they will be, we shall say, vectoriell and generic categories whose refusal would render our words unintelligible.

We can grasp something else in the Cross if, for example, rather than referring it to itself and doubling its materiality in the mirror of its signification, we treat it as a certain indivisible matrix, or at least as an experimental apparatus. It is an experiment carried out in a relatively determinate or closed site, the vessel of Golgotha, with a torturous material technology, no doubt, but one that must prove or quash the hypothesis that Jesus is the son of God, a negative result recognized and uttered by Jesus himself, ultimately a device for the infinite interpretation of this defeat that is called "Christianity," an interminable taking stock while awaiting the return of Christ to purify the accounts of humanity and draw the final line.

In interpreting the Cross, it is essential to no longer think historically, in a determinist or macroscopic manner. It is an indivisible phenomenal bloc. We maintain on the one hand that it is a question of a double death—the death of God through the death of the singular individual Jesus; and on the other hand we do not separate as two historical sequences the Crucifixion, leading to the Tomb, and what is usually called the Resurrection or the "rising" of the Son to the Father, the Ascension. It is a question of thinking these phenomena in the "holistic" spirit of quantum nonseparability, and according to unilateral complementarity. This drama is made of continuous or entangled phases that the theological imaginary tends to dissociate or atomize into moments, according to the corpuscular logic of theology. This is quite understandable: the point of view of each of the two religious contexts and their theology presupposes itself in itself, and hence can only apprehend autonomous denumerable or discernable events, so as then to bind them back together, cause to consequent, according to an extraordinarily trivial and "literary" temporal determinism. Whereas the ensemble of the Cross and the Resurrection interpreted as a complex matrix rather than as a final synthesis or sublation is indeed that which tears the Cross from its barbarous "religious" meaning and underdetermines it as what we call a suspended sacrifice.

As presented at first sight to the eyes of the world, it is a theologico-dialectical folly and an authoritarian cruelty that smacks of ancient barbarism. Christo-centrism is articulated on the crossed—that is to say, the crucifixion as dominant phenomenon—an empirico-historical form of experience that is armed with a torturous technology typical of transcendence, and that is made to function in exteriority according to the theological apparatus. From this point of view, it is a recognition of religious forces that multiply one another or conjugate with one another, and culminate in paradox carried to the point of folly. But this is the sickness of Christianity, not necessarily that of Christ. Christ is crucified first upon the theo-christo-logical or religious Cross formed by the Logos and the Torah. The folly of the Cross is first of all that of these crossed timbers, Logos and Torah multiplied by each other, like the equal variables of transcendence that must be conjugated, or the symbolic parameters, the still-transcendent coordinates of the Crucifixion. We renounce every univocal and continuous interpretation of Christ on the basis of that supposedly central ritual experience and of a mythical circularity; we break this essentially philosophical continuity. It is a question of rediscovering immanent faith produced by the generic matrix that is the nonseparable bringing together of the Cross and the Resurrection against idealizing contradiction and folly. The whole problem relates to transcendence and to the reversibility of the Cross.

Although it is therefore not just a raw material like any other for a science, and although its generic content is manifest in a scientific way, *the Cross is an ideality but a materiel one, a materiel(-)a priori phenomenon, a fusion or superposition of a priori matter and significations less historical or empirical than the traditional ones.* Logos and Torah are ideal structures that function a priori but immanently, in vectoriell and materiel conditions. We must not idealize and separate symbols in a transcendent way and speak of metaphor—this is to conflate the immanent lived of the experience of the Cross with spiritual and metaphysical realities. Without this materiellity, moreover, one can understand nothing of the passion of the mystics, of the imitation of the Cross, nor of the fetishism of the Catholico-pagan Church. We must constitute the a priori and concrete sense of all of these phenomena, such as the descent from the Cross. When materialized, the deathly effect of the Crucifixion will no doubt take on a banal and pictorial form, but one that must be deciphered, not just represented. The body of Christ penetrated by thorns and nails, crucified, arms wide, offered up to the world that he abandons and to the God who has abandoned him, losing his unity as an individual, a wavelike unity of breath and a corpuscular unity of blood that ceases to *circulate* within the interior form of a living creature, to be diffused as life diffuses outside of itself: will he not bleed eternally? Is this not a physical body raised to the state of a fiction-body?

What is "abased" in this operation? To decrucify Christ, for example, taking into account the "descent from the cross," is to admit the existence of a transcendence, but one that falls; it is to submit it to a christic *clinamen* through which God the Father and Jesus the subject are thrown into-immanence—the immanence of the tomb—one ceasing to be a manipulator, the other a victim. The descent from the cross has too often been treated either as a moving pictorial motif or else as a facile symbolic scene. Would the Crucifixion not rather be the consummation of transcendences: the transcendence of God and of theology in his Son, and the correlative transcendence of the individual? It is impossible to isolate the Crucifixion from the Resurrection, and the Resurrection from the Ascension, as evental, "corpuscular," and atomizing sequences, without a return to historical determinism and sacrificial barbarism. It is here that metaphor is facile and dangerous. Is Mount Golgotha a geographically localizable place or a spiritual summit, or is it the form of all lived transcendence, with its amplitude gradient leading up to Calvary, and its fall leading to the tomb as degree zero of transcendence?

THE TRIAL OF THE CROSS AND TEMPORAL DETERMINISM

The Christ-science is not yet constituted; yet it can be divined, as impregnated by theology as it is, as the reciprocal interpretation of two contextual discourses, both of which, however, are destined to be brought down or to decline all the way to the tomb. The Crucifixion is, properly speaking, but the "preparation" phase of a materiel-physical (rather than symbolic-material or metaphorical) experiment, which, according to the historical order, will be followed or completed by other phases. Does not the grain of faith of the "believers" attest to this historical and literary reading as a non-"scientific" drama? We have instead conceived the entire set of these phases as forming a device wherein is prepared a crucial experiment designed to decide between the theological and the generic interpretation of Christ—that is to say, to posit the real or immanent history of Christ as irreducible to Christian theology's account of it, an account with which it is not commutable.

It is here that the historical determinism of theology must be called into question. It transforms the sacrifice of Christ into a "first cause" of the Christianity of the Church. This still-macroscopic image of the Cross in fact only initiates a sort of becoming-Dionysus of the crucified who has not yet attained his threshold of generic quantification. If we admit that the formula of the "converted Jew" translates the reception of Judaism into the terms of Logos, then what would the interpretation of the reception of Logos in Judaic terms be, if not closure within being, a transcendence closed up in the immanence of the tomb? But the process of the Cross cannot stop at this antinomy, otherwise it will see nothing in it but a scene, a theater. Something else however, an unprecedented event, had to be produced virtually or is in the process of under-going: the resurrection of Christ and his ascension, a new amplitude of history. Descent to the Tomb, Resurrection, and Ascension can only be phases of the same event and *must be able to be read, but they precisely are not read*, in the very emptiness of the tomb. If the Cross is understood on the model of what takes place in the generic matrix, it is more complex than an enshrouding rite, and does not belong to the logic of temporal sequences isolated or linked together by the imaginary of a "sense" that would ultimately be in the hands of God. As indivisible or global experiment, it contains a vessel that is closed but capable of being reopened, which tries to "trap" Christ as one traps a particle in the void. Christ is trapped in the void of the Tomb as in a black box—he descends into it alive, he arises from it alive, Christ is dead (and) alive! What has happened, what paradox worthy of Schrödinger, but with an opposite meaning to his? In a sense, we shall never know what happens here, whatever projections we might make of it. And yet there has been no lack of efforts to distribute the events one after another, to establish an intelligible causal order at the price of the unintelligible paradoxes that we try to totalize, war-weary, in "the Mystery of the Cross." The Empty Tomb is an experimental vessel that belongs to the Cross—that is to say, to the process of Resurrection—not a foundation. It would contain, in the state of an axiom, the program of the under-going of the living Christ, were it not for the fact that Christianity hastens to partition this coming and to flatten it onto the "good sense" of a story to tell believers. The Resurrection itself as historical event is threatened if we take it as a historical event via the fundamental axiom of faith; and whereas its importance can hardly be denied, its isolation from the other moments, its institutionalization as the fundamental and founding dogma of faith, risks inverting its futurality and relegating the Ascension to the rank of a terminal effect, a consequence, like the result of a process, and transforming it into an object of fantastical belief. From our point of view, the Ascension, which is often passed over in silence, or is accentuated least, regarded as an effect that follows the Resurrection, is, along with the latter, the more than fundamental (prior-to-first) operation that underdetermines the sacrifice and the obscure night of the Tomb—but why? After all, theology is storytelling.

FROM PHENOMENOLOGY TO THE QUANTUM THEORY OF THE CROSS

The Cross in the narrow sense as the apparatus of crucifixion is part of a vaster set of phenomena whose meaning is indicated by the formula that is coined for it, Christ "dead and risen," but without rendering it intelligible for all that, and not without associating it with an invincible determinist appearance. For, on first sight, the whole scene is laid out like a succession of paintings for an altarpiece—it is a nest of interpretative (particularly symbolic and idealist) possibilities, the intuitive image of a more profound conceptual paradox, the paradox given in flesh and blood by this crossed duality.

To exit from this symbolic, sublimating, and christo-centric vision, we must first take seriously this torture device, perceive it with the intensity of mystics on the verge of hallucination, grasp this material support otherwise than as a wooden support for an operation whose political and spiritual meaning unfolds elsewhere. The wood of the cross is an a priori materiality, as we have said—even its form is invested with an a priori that is not empty, its figure swiftly makes sense of its crossing: it is not only the crossing of a double operation or of a symbolic event. It is also prompt and clean in imitation and repetition. Just as there is a materiel formalism in the generic science of humans, so the Cross is the intimacy of an affect, a formalism, or an operator of materiality: we are tempted to say a nonmathematical algorithm for a stance, rather than a foundational tale or a theater, as certain "believers" have felt it to be.

Let us remark then that two types of hesitation or oscillation are possible in grasping the phenomena of the Cross. The first is macroscopic: interpretation hesitates between intuition and theological concept, image and sense, the two faces of the symbol, one in which sense is rooted or schematized in intuition, the other in which it aims at an ideal sense. We have hardly left the level of a phenomenological or quasi-linguistic apprehension here. The second hesitation is deeper and does not entirely overlap with the rationalist and critical duality of intuition and concept: it is the hesitation between the macroscopic apprehension and another more secret apprehension, still macroscopic, but that opens the way to a quantum apprehension of the phenomenon, and this in two ways: First, on the quantum model of the perception of a cube. The Cross is a whole precisely centered on the central position of Christ, who himself is inscribed in a circle. The pictorial perception of the Cross takes this ambiguity into account, but theological thought seems to take less notice of it. Now, this ambiguous figure is a whole given in transparent perception, but is now convex (the cross brought to the fore or in relief, protruding into the foreground), now concave (the cross receding back into the circle). This oscillation of perception as it hesitates between relief and withdrawal is a phenomenological topos that has taken many forms, some dialectical, but it gives rise to a quantum interpretation once it is a question of explaining in what way the perception of the sense of the figure is aleatory, hesitant, unexpected, and only probable. It is obvious that in ordinary theological perception the Cross is a corpuscular rather than a wavelike or interfering object, which confirms its reading in terms of scandal or folly whose element is transcendence. But it must be taken in all of its phenomenological aspects, in particular, the internal oscillation of its perceived-being. All the more so given that then the conditions are in place for another, nonphenomenological and nontheological mystery to appear in the form of a circularity of the Cross distributed in an entirely internal quadripartite, in four sections or quarter-turns. Here again the paradox is reinforced by the insistence upon the torsion or the crossing of an opening and a closing through which it is reflected in itself, and turns to a cruci-centrism that is perhaps not its "last" word, even if it is the last word of its transparent circle. From this point of view, which initiates a quantum theory but a still-corpuscular one, reduced to its two axes, which break up the circle into its four quarters, it is the bars upon which Christ is crucified or dismembered that appear: the horizontal bar of the Logos, the vertical bar of Judaism.

In other words, what we have here with this oscillation is a quantum temptation, like that which physics faced at the end of the nineteenth century, hesitating between waves and corpuscles. But we no longer have this unilateral duality: the generic and the generic usage of this matrix as immanent are missing. The transparent cube has become the *transparent circle of the Cross*, the specular theo-christo-logical milieu. God and Christ form an ensemble that is transparent through its oscillation, legible now as circle of divinity, now as christic cross. Transparent, concave or convex, the circle of the Cross is the infinite circle of divinity, now centered on Christ, of course. Everything is at stake here, nothing is settled or taken for granted, because we remain in full philosophical and theological positivity at this point. At this moment of perception, the moment of the crucifixion, the Cross no longer has any christic meaning. Above all, it only has its future Christian meaning. It is "in" the circular sense of history, just inhibited by Judaic countertime, and not yet in the sense of the futurality that the resurrected Christ will inaugurate. It would have had to have ceased to conflate or even place in parallel the historical sequence and the essence of the vectoriell event of the Cross, ceased to overlay them onto each other as belief and faith. *To have ceased to conflate the historical object of the Gospels and the object of faith*, the Cross completed and closed upon itself with the Cross vanquished by the Resurrection and under-going prior-to-priority. Rather than inscribing it as it has been inscribed, in the little turbulences of history, in the great circles of philosophy, and finally in the theological cosmos of creation, we now understand the object of faith in terms of another genealogy that would identify in the quadripartite of the Cross the initiation, only the initiation or the symptom, of the quarter-turn—that is, of the Messiah-factor.

Messianity is the real content of the Crucifixion. The real content of the Gospel is not its apparent content: it must be understood according to an algebraic or quartial (rather than "partial") "logic," as an event that is neither total nor partial. It is less a turn or a conversion of thought that is necessitated here than a negative, "impossible," or "irrational" quarter-turn, a futural uni-version or an entry into the sphere of vectoriellity. The quarter-turn is not an abstraction of the theological circle. It is a vector identifiable in the Ascension, which is often the moment that is forsaken, the least visible moment in the christic phenomenon, the least empirical or the least historical. We understand the Cross in terms of the Resurrection and the Ascension, not the inverse, which is the confusion of the real or futural movement of Christ with the apparent movement of Christianity. This is why when we speak of the Cross we do not separate the Crucifixion, the Descent and the Entombment, the Resurrection and the Ascension, treating them as an analytically indivisible phenomenal bloc, but also as synthetically indecomposable in terms of Logos and Torah. And to make messianity and faith entirely intelligible, we must once more consider the vessel that these three moments form, and treat it as a site for the experimentation and the performativity of faith. Otherwise the Cross will remain a philosophical matrix—that is, a matrix that is interpreted twice philosophically, by variables both theological and ultimately religious as well as scientific, and then by the inevitable doubling of philosophy rather than the superposition of science with itself.

CHRIST AS LAST INSTANCE AND UNIVERSION OF THE COURSE OF HISTORY

The establishment of the generic meaning of the Cross as a matrix thus concerns the order of evental phases, not the multiplication of variables by one another. In particular, the Resurrection and the Ascension are more than primary: *they form less an inversion of the course of history than a prior-to-first uni-version or ultimatum*, generic rather than philosophical. *As* vectoriell phenomenon, the Ascension cannot itself enter into an order more powerful than it: the Resurrected does more than invert the empirical and rational sense of history, he uni-verts it quantically and generically, futurally transforming the order as a function of the imaginary quarter-turn.

Thus the Resurrection and the Ascension are noncommutable with the history that follows from them and claims to encompass them. The Real of Christ underdetermines the new order, which is no longer determin-ist, from the past toward the present, but without simply inverting it or denying it, since the Resurrection is futural, and the resurrected Christ the *ultimatum* that under-goes forth to meet unfaithful but "evangelizable" world-subjects. The course of the world is also that of history, not just the simple course of empirical events, as in Hegel, which the Concept confronts. Christ and the event that, without destroying the general course of the world, abases it in respect of its omnipotence, or transforms it by messiah-re-orienting it. Messianity is not an inversion of causality but its underdetermination, and the underdetermination of the historical order. This is why the Ascension is that which underdetermines the Cross and endows it with its futurality, the theatrical or dramatic storytelling aspects being secondary effects of this. But is there a "Christian" catharsis? Would it not be, rather than a purifying spectacle of the subject, the world condemned to relinquish its grip? A fall symmetrical to that of original sin, which was already the responsibility of the world, in which the world would fall from its greatest height? The Resurrection and, consequently, the Crucifixion are profoundly contrary to the spirit of tragedy and sacrifice, which is that of philosophy. The "plan of salvation" as it has been ordered by the sins of man, the sacrifice of the Son and his resurrection, comes down to burdening humans with the sin that is the world, and laying out the crucifying experiment in terms of historical determinism, as an incomprehensible process. Always the planification and the hierarchy of theology rather than discrete and continuous phases. Futurality is a quantum and generic notion, a reestablishment of the temporal order for generic humans. As event of "reconciliation," the Resurrection makes for a rather miserable miracle, an object of belief governed by history and driven by the Churches, a sort of "happy ending." If the miracle exists, it exists in an entirely other temporal economy, in a form as virtual as can be, but it is neither first nor last in the series of history; it is prior-to-first, a stranger to this history but acting within it.

To sum up, the Crucifixion is first from the point of view of experimental appearances and the subject that is implicated in them, but the Ascension and the Resurrection together form a prior-to-priority: more than a "beginning," the very condition of the functioning of the matrix, what we call its Last Instance. This is why Christ is not a new foundation within a history in progress, but the Last Instance that abases history as a point of view. Neither the Resurrection nor the Ascension are objects of belief, because everything in belief is opposed in them, or makes of them miraculous events worthy of "fools for Christ." They are the objects of faith as messianic in-the-last-instance. But together they form an act of emergence that Christianity has understood as source-event, positive or indeed negative certainty, reducing their prior-to-priority to a more banal historical priority. Ultimately, Christ does not invert the course of history, but uni-verts it or includes it in himself, without for all that reconstituting a finality—but only an orientation, cognizance of which, or faith in which, will only ever be "probable."

Twelve

THE SCIENCE OF THE RESURRECTION

FROM RESURRECTION TO MESSIANITY

Supposing that a "science of the resurrection" (Porphyry) is possible, how is it to be imagined and formalized? The Resurrection is the retroactively foundational myth of Christianity. Our project is to draw from it the possibility of a rigorous fiction, a christo-fiction. It is a condensate of Judaic preconceptions and Greek images, which Christianity has brought to an unequaled amplitude. Some of its interpretations are remarkable. Its interpretation can be an act that is variously empirico-imaginary, metaphorical, or mythological (this is the lowest degree of interpretation). Or else a spiritual act, an idealist and mystical interpretation between re-creation and renaissance. Or again, in topological terms of a subject-body incarnating an ideality, a materialist interpretation symmetrical with the above. These interpretations oscillate between the imagery of popular belief, the dialectic of conceptual sublation (Hegel), and topology. The interpretation proposed here intersects with all the others without being identical to them; it is said in human terms, "in-the-last-humanity," in terms of lived materiality, but not in materialist terms; in terms of algebraic objectivity but not topological objectivity; and finally in dynamic terms of vectoriellity rather than static and corpuscular terms. This interpretation posits a concentrated hylomorphism in the state of fusion: it is not humanist but generic, not macroscopic but quantic, it does not make of resurrection an operation of dialectical sublation, but an insurrectional and generic event. We have to come back to the simplest notions. We lack a scientific, that is to say, quantum-theoretical, theory of death and resurrection. To resurrect can be said in Greek as either "to awaken," "to rise up," or "to recover." It is a question, in these tiny nuances, of the difference between a vectoriell interpretation and a dialectical interpretation ultimately founded on the ontology of being and nothingness. The vectoriell is opposed to the idea of *absolute* death, nothingness-death, the tomb-as-absolute-void, and ultimately to the work of the negative that is the substance of the abstraction of these theories. It conceives it as *radical* death or "sickness unto death" (Kierkegaard), that from which Lazarus is rescued by Jesus's ordering him to arise. Death must therefore be a less abyssal ground than philosophers (Hegel, Heidegger, Badiou) postulate—less an empty tomb than a crypt, a "tomb-as-void" inhabited by cadavers, occupied by victimized bodies. As to the awakening, rising, recovering, there is obviously very little to distinguish it from the gnostic or Christian awakening, or even revival, and from the Greek exit from the sleep of the dead. Between the sublation of concepts, the relieving of a guard, the relay of a race, the raising of children, or the removal of bodies, all of these amphibolies are possible. All these nuances have in common the form of an ascension, but only one has the vectoriell form and the simplicity of a phase, of a nonecstatic ascension as simple initiation of a transcendence that does not close, or of an open movement that can close up or sum over itself but not in itself. It must be recognized that there is something in this equivocity that has made philosophers and theologians rejoice in their common fetishist obsession, and not only because of the duality of operations of *Aufhebung*, maker of miracles and philosophical angels, phantoms, specters, diverse appearings. What we call the vectoriell or quantum interpretation can only reduce this variety to a dynamic factor, a transcending that on the one hand would be its essence or concentrate, and on the other hand will not be closed into an all or into a transcendent entity such as a thing-in-itself, a world, or a saturated divinity. This essence of all transcendence can only be a vector capable of being added continually to itself or to others, and of being multiplied without fundamentally changing in nature or genus, without itself sinking into a tomblike autoclosure. The vectoriell interpretation of the Resurrection finds what is dynamic in this operation, what is ascendant without ecstasis ($\sqrt{-1}$), that is to say, messianic.

The messianity of the messiah-factor is the radical origin of the Resurrection, the *Ur-transcendenz* that is no longer the transcendence of "Dasein" nor of the "Ur" whence . . . Abraham set out.

Messianity is what there is of the continuous, if not the repeatable, in the subjective raising, in the emergence or the "relief" on the basis not of the bottom (*fond*) of the empty tomb but of the void as the *resource* (*fonds*) of the tomb. The ascent or the insurrection of the Resurrection is measured by the phase that separates it from the fund of the lived that is inert or deprived of reprise, and not by nothingness, which would never inhabit a tomb. We come back to the inert and "dead" tomb, but not nothingness, as is the belief of those creationists, theologians, and philosophers. The Resurrection is not a creation but a reprise or a resumption of the wave of the lived. The Ascension is the phenomenal meaning of the Resurrection, the vectoriell kernel of every raising and relief before it is closed up into itself or sinks into the enclosure of a historically saturated event. In other words, the Resurrection has no empirically imaginary or factual meaning; it is not even an event closed onto itself, but a transfinite process of phases. It is not even an event that brings itself about, except through its idempotent reprise, because it is the very-first beginning of the event, that which, with the flash of the Logos, exits from the tomb of the night, and the radicality of whose emergence makes for christophany. The Resurrection is certainly completed or terminated at every instant, but it continues in us without closing or being closed. It is not eternal like a current present: we shall say that it is futural.

However a correction, or another accent, must be added to our first impression of these rather precipitate statements. As the Gospels say, Christ is not seen or visible through an operation of subjects. *He makes himself seen.* And that which makes itself seen is less the Resurrection itself than the Resurrected in-person, if we might say so. We must accept that the Resurrected is a being that is doubtless "saturated," but saturated like a clone, perhaps, like a christophanic figure that is perhaps particulate and not corpuscular, in a vectoriell state of emergence and ascendance. For very obviously the Resurrection follows the Resurrected and not the other way around; it is not an operation carried out upon a certain Jesus who has just been crucified between two thieves; the Resurrected underdetermines the images and concepts of Resurrection, whatever religious sphere they come from. This is why

Resurrection is an operation that develops in a milieu of quantum but generic, vectoriell rather than spontaneously philosophical, objectivity. Fidelity does not consist in seeing so as to believe, nor in believing so as to see (we are no longer in the apophantic context of Greek representation), but in being subjects filled by the superposition of the image of Christ rather than being filled by an intentional act. The Resurrection is generic, and thus can be individual, but the inverse is not the case, and within these limits one can say that the faithful are "seers," "voyands" in the sense that psychoanalysis speaks of "analysands." Christ resurrected is more of an icon than an image perceived according to the supposition of the world—the noema of a generic, nonindividual intention.

VECTORIELL SUPERPOSITION AS ESSENCE OF ASCENSION

To begin to make the Resurrection and the Ascension minimally intelligible, we must distinguish carefully between the transcendence of opposed and mixed discourses that harbor an unintelligibility for the faithful and that are "finally" brought down, and the vectoriell "ascending" of the Son, which is incomparable with the image that philosophy gives of him, and which can only be the superposition of Father and Son, with the death of the authoritarian, persecuting, and sacrificing Father. With the empty Tomb and the Ascension, the scene of contrary interpretations is thus terminated or destroyed. The Ascension retroactively gives its meaning to the scene of the Crucifixion, and it is not an idealization like the "Ascension of Hegel," as Marx says mockingly, or a return of Christ "alongside" his unchanged Father, an entirely theatrical, even comical scene. We oppose to it a materiel (rather than idealized or metaphorical) reading of this scene of the Cross—a christo-fiction, perhaps. Thus the greatest prudence is necessary in the cry "Christ is risen!" which responds to the declaration of Nietzsche's madman, "God is dead, we have killed him!" The inclusion of subjects in this crime is most interesting. Unfortunately, the joy of the proclamation stifles the minimum of reflection that is necessary for this affirmation of the new faith. The popular belief in spirits, relayed by the silence of theologians avoiding having to respond to this philosophically impossible question on *the real content of the Resurrection*, has done the rest. Its mystery adds in an almost incommensurable manner to the enigma of the death of God; it has remained a slogan or a proclamation, at best a confession of faith. But we can be sure that the two formulae that enframe the Cross have the theoretical coherence of a system that apparently has the ring of a dialectic, toward which philosophy is often led. However, just as one can only kill God under certain theoretical conditions, in general replaced by reasons that are contingent upon history, Christ can only be "resurrected" under equally drastic theoretical conditions. We must be able to say not "Christ is risen, we have resurrected him!," but "we are resurrected in-Christ." This formula is obviously a nest of aporias and interpretations, just like the death of God, and risks remaining especially unintelligible. And yet, under pain of reconstituting a simple pagan belief, we cannot excuse man of all responsibility, unless we think according to philosophical belief the real process, which is that of a unilateral complementarity or an indivisibility of the Crucifixion and the Resurrection. The risk is once more that of a vicious circle, since faith is the expression of the superposition of Father and Son under the Son in the Ascension—as if the idempotence of the Son replaced the omnipotence of the Father. The matrix as Cross and Resurrection is ultimately inclined generically, and underdetermines spontaneous belief rather than determining it in a rigid or circular mechanism. Messianity is the *generic condition* of the beginning of the process of faith, but not the beginning itself, which is why it is virtual or is a nonacting that underdetermines the occasions of acting—that is to say, the occasions of its immanent practice.

INSURRECTION RATHER THAN EXISTENCE

What is the effect of insurrection upon macroscopic bodies, what is their under-going as resurrected? Insurrection is a vectoriell phenomenon, a lived materiality structured algebraically, or again, more intuitively, an ascending or a nascent wavelike phenomenon. And it belongs to the wavelike to be an ascending without objective ecstasis. Insurrection is prior-to-first and places under condition the historical order of resurrection or revolution that would be primary. This is the task of the faithful: to raise up the world like a wave, and to exhale it like a breath. Revolutions are the way in which history breathes, and it often dies of suffocation.

Messianity has never been a more or less direct diachrony, but always a raising-up whose effect is to reduce dominations and to lead them to their fall into immanence. Like the linearity of the trace and of the diachrony that it still thrives on, double transcendence is forced to precipitate in the simple transcendence of a clone or a faithful.

Philosophy conflates the ascending of immanence with the theological double transcending that covers or envelopes it and that is decomposed into a clone. The vectoriell model allows us to think pure raising-up without referring it to a site—it is not even the event as part of a named site, it is no more the outbreath of an e-vent than it is an ex-istence, the ex- that is a repetition of the ex nihilo of creation, a theological residue. Surrection is an ascendance or a phase that no longer goes to trans-cendence, the emergence of ex-istence insofar as it does not go to the ecstasis that loses itself in the object. The more existence closes up in itself and is saturated with the world, the more raising-up is not affirmation of existence but its bringing down. Raising-up is that which under-goes under the raising, which raises itself up without rising. Raising-up is neither ontological nor existential but human and generic; in any case, it is that which underdetermines and does not sublate. We shall not even speak of ex-surrection—for this would still be to think on the basis of nothingness, of coming back to the soil or to the earth from which the philosophical tree emerges—but of resurrection as in-surrection. No creation of an exteriority by the throwing of immanence that is content to in-surrect in-itself.

The immanence of the wave *as simple ascending* is not seen as the particle is seen; the angle of its phase or its inapparent thickness will only appear later with the fall of transcendence, and moreover will appear *as inverted verticality*, like the fall of the great idol. The fall of double transcendence is fundamental here as the effect of surrection. Messianity is an intentionality in reverse or against the grain, which displaces the philosophical subject by reducing it to the state of a clone in-immanence. The vectoriell ascent of immanence and the descent of corpuscular transcendence not in itself but in-immanence—these two phenomena go together. The Last Instance thus does not sublate transcendence one more time, does not mediate it, does not conserve it in the form of traces or interiorized phantoms. Its effect is the fall of the monotheisms that it underdetermines.

As to the Resurrection supposedly fulfilled in mythological or theological manner, taken as term or corpuscular entity, it flattens the wave and its superpositions with the soma-body, flattens the lived with the world, and takes a point of view called "macroscopic" or decoherent; this is to conflate the resurrected—that is, insurgent—body with the transcendent body of the world, living or dead, it matters little: "Arise!" Not "rise again," like a bad repetition in place of the insurrectional reprise—such an injunction presupposes that one was standing, was then dead, then resurrected. Resurrection then risks being an ideal repetition of life if insurrection is not a reprise resuming the incessance of the lived. The generic body is philosophically indivisible between soul, spirit, body (soma), which are residues of the transcendent or "decoherent" divisions once their superpositions are destroyed. As to the "flesh" that is from the start "incarnation," it is this nonseparability and this nonlocality of the clone as particle, borne or contributed by the wave of the lived. In other words, faith is the flesh itself qua correlate of messianity and underdetermined by it. The glorious body and the faithful flesh are one and the same.

SCHRÖDINGER'S LIVING AND DEAD CHRIST

Resurrection and Ascension are phenomena of materiality, and find their ultimate explanation in an ascending, an essentially vectoriell act of wavelike superposition that remains the Same as idempotent. Obviously this is not to suppress what there is of the "miraculous" in them, because the notion of an idempotent wavelike ascending, strong analytic and weak synthetic, is just as mysterious, but just as natural, as the algebra of quantum theory. Nonecstatic Ascension corresponds to the algebraic property of idempotence and it is a pure Surrection, which, like every scientific law, in a sense has no justification other than itself—it is ultimately axiomatic. The last "foundation" of the Ascension as Resurrection and of the Resurrection necessarily as Ascension is simply of an oraxiomatic order, or of-the-last-instance, and belongs to a strange logic of idempotence. If one is surprised at such a "materielist" and formal reduction of the Resurrection to a flat algebraic property, why not be just as surprised at that marvelous property of algebra that, once materielized, prescribes that every immanence is ascendance? That if there is a plane of immanence, it ascends, doubtless because of its nature, which is vectoriell or angular, and subjects it to a phase? This undermines the philosophical logic of identity with vectoriellity. Insofar as it is not a "reduction" but a fusion or a cloning by idempotence of the lived taken from the human subject, this is a question of materiellity and not materialism, of a "real" rather than an empty formalism. Will we go so far as to say that the experience as such of the Resurrection is for the faithful an experience of cloning? Ascension and Resurrection are *prior-to-first*, this means to say that they do not determine themselves circularly, but underdetermine themselves, that they are the real condition of the Cross but that they themselves already suppose such a fusion. It may be that in order to think cloning one must conceive it as both operator and operated-on, but operated-on in-the-last-instance. And the theory of the Last Instance destroys the vicious circle or resolves it through the superposition or "resumption" of the matrix of the Cross.

The Resurrection is not a new creation or its repetition, but the very-first time where precisely there is finally a creation in this world that is not a repetition of a worldly procedure or of a phenomenon fallen back on thought. The gnostics began to think most rigorously but still religiously, they thought the divine creation as a failure; another was necessary, through Christ, in order to finally succeed where a failed God had been unsuccessful. But for quantum gnostics, there has never been a creation of the world or in the world—it is the world that is "wicked" or "evil," and consequently also the God who claimed to have created it and yet hesitates to assume it. But Christ is the prior-to-first creation of novelty who takes the order of things the way it should be taken—through a futurality that is not of the order of time, except insofar as it haunts it as a Stranger, the only one who is not Eleatic. It is a question of establishing the order, not even of reestablishing or remaking it in its original, supposedly lost figure. And the determination of the order or the Last Instance is the Son who rises or ascends and brings down the Father. The Ascension is just capable of superposition; radical immanence is its work; above all it does not form a plane of immanence like a secularized form of the plan of salvation. It is prior-to-first in relation to the Cross that nevertheless comes first in the evangelical story.

As for the Tomb, it receives he who will ever be a living being and who will now receive the lived-without-life. Although empty, it is a tomb, not a cavity dug in the soil, but empty because reduced to a positive o, a o that is not nothing. It is still life, but in a generic mode, not a life idealized or reduced to memory and its disappearance: the quantum problematic prohibits that idealization. The Ascension at once brings with it and abandons the cadaver (hence the cortege of simulated and phantomatic reappearances of Christ). If we contract, as indivisible, the phases of the Gospel stories, we obtain Schrödinger's Christ, with the Resurrection and his cadaver superposed. But the Ascension is more intimately still under condition of the Resurrection insofar as *it resumes the lived as subtracted from life*, an absence that is not an idealization. The empty Tomb = o represents what we call the lived-without-life, delivered from the human subject and from the individual body. The lived itself is not a generalization attributed to a quantity of nonhuman so-called living phenomena; it is distinct from any life whatsoever, it is of the lived in the generic and thus human sense. The human is living, no doubt, but is life rendered human, as Marx said, and also death rendered human—that is to say, lived. And if the lived materializes ascending, the latter as superposition is the very form of the lived, and distinguishes "life" rendered human or generic from other forms of life.

As for the Resurrection, it is indeed a resurrection of "bodies," which is a way of opposing it to all Greek idealism, although it is still a matter of a reading in terms of the world rather than in terms of the Ascension. It is certainly not the separation of the soul-as-Idea and the material body, nor even the doubling of a spiritual body and a worldly body, like the generation of phantoms or the living-dead. It is a separation of the abandoned life of the world and the lived that is raised up from it and superposed with itself for reasons that we know are now or never of a scientific or quantum order. Only the lived-without-life grasped by idempotence can ascend or make for ascension. Life is dead, the lived is resurrected, not re-created but separated from life-death. This undoes the unitary confusions and amphibolies of life and the lived. And this is why the concept that speaks best to this reprise or superposition of the lived with itself is "resurrection," on the condition that we do not understand it via common sense and belief, any more than we should understand resumption as repetition or superposition as a new positing. "Re-surrection" is not a repetition of a living being or of creation, nor even of life, but a re-sumption of the lived, an insurrection of the generic Christ out of the world. Christ is the insurgent of the Tomb of the world.

THE ASCENSION OF CHRIST AND THE ABASEMENT OF GOD

We have distinguished the simple and transfinite Ascending of messianity and the ascending completed in itself or the double transcending that is always, in every theology, a double movement in opposite directions. Now, the effect of vectoriell Ascending is to abase transcendence in an entirely new way and definitively, reducing it to a simple or particulate form. The Ascension is thus also the abasement of the divinity (which theology supposed absolute) of Christ. It gives the Cross its meaning, renders it necessary as the bringing down of God. The Ascension of Christ does not suppress the divine dimension of Christ that would substitute itself for God or take God's place, but abases this dimension or subtracts it from its specular image, thus destroying the theo-christo-logical doublet. *Christ is the unilateral complementarity of messianic ascendancy and the abasement of the belief that besieges the faith of the faithful.* The Ascension is the christic amplitude of thought, not an external, marvelous event. There is no dialectic of Ascension and of Abasement, but an ascending so as no longer to transcend indefinitely.

We have evoked a certain factor of a scientific order: the messiah-factor or factor of christo-fiction, capable of subtracting the lived from the order of life and making of it a generic subject. What is it that ascends without doubling, that superposes without doublet, and that destroys doublets where there are any? That which can ascend while superposing itself or adding itself to itself but remaining the Same is messianity as a phenomena that is vectoriell rather than positively vectorial or geometrical, and that must therefore be materielized, rather than schematized, in some human lived. But to understand vectoriellity we know that we must go all the way back to the quantum-theoretical understanding of it as quarter-turn or $\sqrt{-1}$. We perhaps understand better that the real or christic order does not "begin" historically with the immanent Ascension, but is nevertheless underdetermined by it, by that act of superposition that is valid for vectors and more profoundly for those materiel entities known as "complex." From the quantum perspective, the Ascension is the superposition of Father and Son as vectoriell entities, or more exactly the superposition of the Son as vector of the quarter-turn and of the Father reduced to the Son, with what remains of the divine in the Resurrected—that is to say, generic messianity. Why, in vulgar terms, does the Son not engender the Father? Or more scientifically, why would the generic community of Children not be the immanent deduction from the life of the Father?

If we wish to think this in a quasi-theological way, here the negative or subtracted quarter-turn as $\sqrt{-1}$ must draw on the Logos and the Torah: with Greek immanence as its modulus aspect and Judaic transcendence as its phase aspect, aspects that become indiscernible in Christ as this always-resumed vector. Why, in the historical story, does the quarter-turn have no particular, genetic, or vectoriell function? Why does the Ascension disappear as a fundamental theoretical moment between the Crucifixion and the supposedly completed Resurrection? Precisely because the theological reading is historical, and is traced from the course of time or the course of the world, which traverses the circle of the Cross as already given positively, a circle repeated in completed and closed sacrifices, susceptible of repetition. But the negative or subtracted quarter, as $\sqrt{-1}$, defines the vectoriellity or the wave function that takes place outside of circular space, endowing it with its lived-without-life materiality. It alone can define the state vector of the Christ-event and rectify, include, and condition the course of the world in the futural direction of salvation. The quarter-turn brings into play a vector as module and phase. In relation to theological representation it is an "imaginary" and complex vector that is situated in a space other than the perceived and the geometrical, already algebraic but not yet reduced to generic immanence. However in theology, above all determinist theology, Aristotelian in the oldest cases, Newtonian in the most modern, positive geometry has been bolstered by the contribution of the Parmenidean conception of the Same as specular circularity. The quarter-turn as "impossible" dimension breaks with this object that is perceived, and above all frozen in its perceived-being, encysted in itself. The undermining of the circle of the world and its autoperception begins precisely with the replacement of the convexity of the circle by the convex cross: putting the Cross into convex relief necessitates the division into quarters, the first operation that announces the undermining of the old world and of the ancient archaic forms of sacrifice. But when perceived as convex or bulging, a fortiori as concave or recessed, the Cross is still a Greco-Judaic mixture dominating Christianity. When the quarter is finally understood in quantum terms, and no longer as convex but as itself concave or as $\sqrt{-1}$, *when the theological power of the Cross is itself annihilated, the Cross brought down and the theological circle with it*, deprived of its positive and deadly power as great object of belief, Christ will have accomplished his under-going, completed his migration out of the bosom of the Father. Messianity and Fidelity will have triumphed. What will remain, as the objective appearances of belief and theology, will be the sinister face-to-face of the ancient figures of authoritarian and barbarous domination that are God and the Cross of his Son as the Calvary he himself prepared.

THE MEDIATOR-IN-PERSON OR
THE MEDIATE-WITHOUT-MEDIATION

Since Christ is a theoretical event without any classically religious or theological message, in order to give this event the message-form or the communication-form it was necessary to go back to the Greek and Jewish contexts of thought that were already formed and spontaneously presented themselves to be rendered "compossible" in their philosophical way—or to have recognized the great work of Paul and the theologian philosophers ancient and modern, the Greek Fathers in particular. It is thus that Christ became mediation, that Greek reading of Christ that we have had to reduce and transform into an axiom of mediate-without-mediation that is the generic itself. In the historical order there is always a time lag between the event and its entry into retroactive discourse in the field of the past. We had to place these discourses back into immanence, transforming them and making use of them as a function of the "instant" of Christ that is once-each-time.

One can see in Christianity a new foundation for human sciences (René Girard) on the basis of sacrifice, which annuls mimetic rivalry. Our interpretation is more quantum than sacrificial, and posits the radical primacy of Christ over a Christianity with which he is not commutable. This is the quantum and thus scientific principle of superposition, which annuls the old mimetic rivalry and assures the primacy of messianity over sacrifice. The emergence of Christ is the emergence of a necessary constant for the theological and human sciences; the messianic constant establishes the sciences of humans upon a "terrain" other than that of the world, on the basis of the last-instance of generic man. The principle of superposition contains the sacrifice of the operation of mediation and extends the messianic suspension to every religion, Christianity included—an explanation, and a consequence, that we distinguish from René Girard's thesis. The sacrifice that annuls the combat of opposites produces not an intersubjectivity of atomic entities, but an interference of fluxes of lived, or an interreferential subjectivity. It is the christic invention not of the dialectical body but of the generic body "in my name."

The principle that governs and orders mimetic duality, the ancient God of sacrifice, is sacrificed insofar as it is *declined* or brought down, but not reduced to nothingness. Becoming immanent in his Son, he is the prior-to-first of every instance as idempotent superposition of messianity, so that mediation as operation is flattened onto one term (the Messiah) and conflated with him, the Mediator. He is generic, universally valid as Same, not as the modality of All. God is sacrificed at least qua separate term: the Father is made Son, he was not sent as a request for pardon or as a message distinct from its messenger, like some kind of spokesman. The generic event does not resolve the problem of diplomacy, of "representation" (we couldn't say it better) or a Jewish quarrel between God and men. The superposition of science and the lived only affects the ancient transcendent God and the equally transcendent subjects who accompany him. From one to the other, with one and the others, messianity is born as mediate-without-mediation.

THE DEFENSE OF HUMANS AND THE SACRIFICE OF GOD

The new Law is underway and at work in each of us, and as such, each of us is messiah. Messianity is in under-going to accomplish the Law in the person of Christ, and not externally. It incarnates the Law in faith, not as an external complement but as a separation of the mélange of faith and belief. God therefore takes on a new function: he is the mask that dissimulates the possibility of a science of religions, he prevents it from being able to establish itself. In his own way, he is an Evil Demon who would make science impossible and allow the psycho-religious imaginary to run free. In short, we must choose: either the religions, including Christianity, to which we give our usual belief, or Christ and faithful subjects; and this is not Kierkegaard's "either . . . or."

We arrive at the non-Christian subject-science through the sacrifice of divine transcendence, the sacrifice of God who, on the cross, abandons his transcendence by superposing himself initially, and thus as positively as possible, onto humans. There is a positivity and a materiality of crucifixion: it is the sacrifice of a God who would symbolize the metaphysical science and who from now on makes way for the christic science and its new constant inscribed in idempotence. It is not Christ who sacrifices himself, any more than it is God himself, who no doubt may well have considered suicide in contemplating his disastrous work. This sacrificial aspect is the effect of abasement or depotentialization by superposition—that is to say, of the transfinite under-going of the resurrected Christ. Is God "dead"? Transformed, rather, having lost his status as the great manipulator, but not because of the world and nihilist society, with which he reaches quite an understanding, and with which he can make contracts and concordats. A science of humans was invented, a generic constant discovered that took God's place as a constant and shattered the theo-christo-logical doublet. Every science sacrifices its phantasms and the imaginary of its objects and procedures. Christ is now the name of a constant defined in the context of his field of objects available to "human" sciences. For our part, we see here nothing of a scientistic ideology; on the contrary it is for us testimony to the inevitable "historical" contingency of every thought oriented or dedicated to the sole imperative defense of humans. When Christ says that he has come to fulfill the Law, it is almost a banality: he cannot but transform a constant in order to propose another, fulfilling the Law as a new scientific procedure, incarnating it or subjectivating it in a more human way. And this generic constant would imply the "sacrificial suspension" of transcendence, God or Other. The "sacrifice of Christ" is a provisionally arrested representation of science but one that, as arrested or blocked as it may be, spontaneously proliferates and distributes itself through theological thought. We should not imagine that science can give birth to continual or repeated sacrificial acts of the type that religions thrive on. Certainly there is no sacrifice of sacrifice as a negation of negation, but only superposition or resumption of sacrifice in messianic immanence. We might call this the transcendence-in-immanence of sacrifice. To take up a term we know to be equivocal, the "sacrifice" of God allows humans to incarnate, in turn, the constant of the Law. This sacrifice of the old God—the gnostic type of interpretation in the encounter with

Judaism—calls for another concept so as to understand that it allows a reinterpretation of the Law by freeing man as generic subject from the constant. God being dead or rather having become sterile, ineffective, or suspended, the constant can finally be superposed with the human lived, which ceases to be soaked up and impassioned by the jealous God. Christ is the superposition of the Law with the lived of subjects par excellence—this fulfillment of the Law has nothing to do with a dialectical operation, it is even profoundly antidialectical. On the basis of this, all men are equal, but generically; this is not an abstract equality like Pauline universality—they are equal in-the-last-humanity as generic subjects *in* science.

THE DESTRUCTION OF THE THEO-CHRISTO-LOGICAL DOUBLET AND THE END OF MIMETIC RIVALRY

The Jewish God is sacrificed for the under-going of the Son with whom he forms the Mediate, he who refused mediation and who, in this sense, was the opposite of the Greek gods. God come to earth, he is sacrificed in his transcendence or his separate and Jewish being. The Christ-science is obtained by using at once two discourses as superposed, Mediation and the Law, and this is the new Law as messianity becoming mediate-without-mediation. It is not an external, substantial, and synthetic combination, as the Church believes.

He can only give himself to the world rather than re-create it a second time if he makes himself idempotence, power of the Same. The Same is a simplified, binary, or dual repetition, a resumption or an overflowing in immanence. He fulfills the Law, meaning that as Son and God he superposes the Father with humans—but in order to carry out this operation, God must be sacrificed in favor of the Son, and must bend to the law of the immanence that science requires.

It is therefore the Father who is crucified, but there is no reason for the Son himself, once "fulfilled," to be sacrificed: he is the one in which the mediation is consummated, the Mediate as hard and indivisible kernel of mediation. The Son is the indivisibility of the God he assumes as idempotence. He is neither sacrificed by the Father nor transformed like the world; he is the unilateral complementarity of messianity and faith, the generic duality made of two events unequal in nature and in function.

The Son is the life that does not die—that is to say, the lived carried to the infinite and to eternity, but on the condition that God "dies" on the Cross as far as possible. God is dead so that generic man may accede to the lived, not to individual man or ego, which die along with God and his mythology.

Sacrifice thus annuls the mimetic rivalry that applies to the relations of God and Christ in the theo-christo-logical doublet, and not only those between humans. *The interpretation of sacrifice must be scientific rather than "sacrificial" or autointerpretive.* The principle of superposition contains sacrifice or the bringing down of its operation of mediation, which would suppose the independence of the first term of the doublet—that is, God. Sacrifice annuls the struggle between opposites, and produces not an intersubjectivity of atomic or egological entities, but a simple messianity fabricated in the interference that is fidelity. Mediation as an operation is flattened onto one term (the messiah-subject) and is conflated with it; it is immanence as the Mid-(of mediate)-without-mediation, in the sense that it is the latter and has no need of another term.

THEORY OF SUSPENDED SACRIFICE

A classical aspect of Christ is obviously of the order that we might call "sacrificial," but what is the relation, exactly, between Christ and sacrifice? There have been too many interpretations going in the direction of the anthropological and archaic-religious imaginary for one not to be suspicious of this doxa.

Christ is the generic incarnation or the prior-to-first ultimation whose theoretical conditions we have already posited, among which are idempotence (the suspended sacrifice of Christ, the immanent completion, and the mediate-without-mediation). Now, idempotence includes a suspension, an immanent neutralization that belongs to it properly and that has the form of an abasement of transcendence. It is the condition, let us say for the moment the "sacrificial" condition, of the possibility of superposing the two Laws. Who is sacrificed? We have posited as an axiom that it is the function of God as Father that is sacrificed in Christ. It is not the son who is sacrificed, there is no reason for him to be—it would be a pure injustice, and the salvation of the world could not be acquired by way of an injustice. The sacrifice that has a true symbolic and foundational purport is not that of Jesus—even if a phenomenon of this type did take place, one that answered to a logic of the Roman State—but that of God on the Cross, of God "in" his Son if you like, not at all of the Son whose Father decides, aberrantly and cruelly, to sacrifice, as is often thought by religions that invert the process. The "sacrifice" of God is the condition for messianity or the christic flux to be able to constitute itself and to be valid for Jews and Pagans. If Christ is indivisible as generic immanence, it is impossible to discern in him in classical philosophical and atomist manner the components of "his" sacrifice. This sacrifice took place necessarily in Christ and constitutes him. But this does not at all mean that Christ "himself"—identified improperly with persons like God or Jesus, whereas he is the generic "Same"—is the sacrificed, and that we are involved in a bad dialectical fluidification of contraries, or even of substances.

Superposition is a sacrifice, if one insists on this term, but sacrifice as unilateral suspension of reciprocal mediation, of the exchange with God. God is suspended idempotently, or becomes Son. What is sacrificed is reciprocity or rivalry, the divine side of mediation, but only on the condition that God should be radically immanent, or should cease to be mediatizable. Two solutions must therefore be excluded.

1. The celebrated "death of God" of modern theologies is a metaphysical symptom of idempotent suspension. The suspension of the divine in favor of the engendering of Christ has almost nothing to do with modern metaphysics's act of killing God, as if God were sacrificed in favor of the Greek law alone. The moral God does not die here, nor does the subject that corresponds to him just symmetrically disappear: it subsists as obedient to the new generic Law in which and through which it is transformed. The abasement of God here is not a belated theme of Christian modernity. It is the condition that is necessary for Jesus to acquire a dimension, a generic figure valid for Greek and Jew alike. The generic duality brings together God, having unilaterally sacrificed his transcendence, and the subject freed from the servitude of the Law but still (under)determined as obedient. God sacrificing his Son is a Jewish image of sacrifice; the Son killing the Father and sublimating him all the more is a Greek image of sacrifice; the Father sacrificed or abased so that the Son may be born—this is the gnostic but not necessarily "religious," and still less familialist,

truth. Christ signifies that there is a veritable immanent genesis of the Son, rather than those all too vicious philosophical geneses that are "family histories," where the family is already present and anticipates the birth and the education of children.

2. Now it is difficult or dubious, as we have said, to speak of the "sacrifice" of the Cross. Here sacrifice takes the form of a decline as an abasement of divine transcendence. Still, the general structure of mediation demands greater precision. *For divine transcendence has itself already been acquired by sacrifice, a sacrifice has in reality already taken place—this is the logic of double transcendence.* Religions are devices with at least three terms, more or less objective and "wrought" transcendental structures. Now, there is always some transcendence, even in the transcendental forms of religious mediation. Here we find, once again, in relation to sacrifice, the doublet structure of theology. It is thus possible to explain that the term "transcendent(al)" is excluded from duality, like a scapegoat third term (*tiers-émissaire*) (Girard), an exclusion that permits the hierarchical organization of the duality and the constitution of a closed group. The transcendental always fulfills a negative and positive function of this type, its exclusion is constitutive of the group that it stifles and hierarchizes. Every religion therefore belongs to a sacrificial activity; but, to be precise, it is not certain whether the Cross, seen in terms of the Resurrection, corresponds *really* to such a phenomenon.

Christ, contrary to what Christianity reflects of him, does not tarry with this old, sufficient model of sacrifice. Current and theological Christianity, if that is what we are considering, does in fact trace divine transcendence from this implicit transcendental model; and, inversely, philosophy imitates religion by transcendentalizing it. But as for Christ as faithful scientific stance, the suspension of his double transcendence cannot be either religious or transcendental. *It is now this structure of sacrifice that founded the transcendental and divine transcendence together that is suspended.* A sacrifice was necessary to produce the transcendent/transcendental God, to constitute a religious community; but now we need an entirely other operation to suspend this device itself. This suspense can no longer be a sufficient sacrifice, even if Christianity would willingly content itself with this solution, thus falling back into pagan errors and vainglory. It must be a superposition.

Because Christ is the "Good God," the ancient God does not need to be "sacrificed," even in the metaphorized sense of a bloody religious rite, but only reduced or abased. The meaning of the sacrificial motif must be transformed, its imaginary depotentialized. The "emissary" God (*Dieu* «*émissaire*»), as we have said, resulted from a barbarous and religious sacrifice, a jealous and evil God, excluded and full of hatred. This primary sacrifice produces a transcendent(al) hatred, whereas the immanental reduction suspends the transcendental itself or depotentializes sacrifice. The sacrifice of the man "Christ" as it is thought in general is a way of accepting the exploitation by the third emissary (*tiers émissaire*), the transcendental organizer. Gnostic salvation as reduction to immanence is designed for the wicked, evil demons and other products of sacrifice, which benefit from it or exploit humans.

The appearance owes to the fact that every sacrifice produces some transcendence that has an aspect or an effect of (ultimately transcendental) immanence. But this is not to do with an immanental reduction by superposition, which is the affair of (and the good news of) Christ alone. God must be brought down, his sacrificial origin itself suspended—this is all that science can demand of religion and of him, not his metaphysical "death" nor his atheist and materialist refusal. The suspense or decline of sacrifice has been confused with sacrifice itself, as if it were a sacrifice of sacrifice, through a specular mirror effect. Christ exonerates victims, but he does more: he suspends the supposed bloody sacrifice of the Cross or shows its appearance. He exonerates the cruelty of the Jewish God by bringing him down from Sinai, to Golgotha. If the first sacrifice founds the community and stretches it all the way to the divine, Christ contracts the latter or superposes it, giving it a generic consistency. The more sacrifice is the affair of priests and of the anonymous and transcendent Law, or of transcendental philosophers who try to soften it up, the more its immanental suspension is the affair of Messiahs and of the milieu of existence of the Faithful. *Strictly speaking, we could say that messianity is the immanence of sacrifice, or immanent and nonreligious sacrifice.* The consequence is that the sacrifice of the third emissary founds a proselytism of heaven and earth, a militant and harassing proselytism, whereas messianity founds only an infinite defense of humans and nothing more. It is urgent, in any case, to pass from the conquering Christianity which is that of the "return" of Christ to a Christianity of the defense of the faithful in-the-last-instance against all religion. There will be no return, nor even a turn, of Christ to make him a generic subject or messiah. Our faith is eternal and actual, it suffers nothing of the passing and contingent nature of our beliefs.

HOW GNOSTICS RECEIVE THE BLOOD OF CHRIST

We do not need to decide, for our account, between a historical Jesus, who we well know was Jewish, and a Jesus become "Christian" and the involuntary founder of Churches—this is not our concern. However, gnosis appears to offer us another perspective that has long been subject to persecution. But it contains an ambiguity that is Platonic in origin. It evidently bears the trace of a conflict that will be treated by and will find a solution through quantum theory: the conflict between two types of representation, the wave of messianity and the corpuscle of belief. In its still mythological religious representation, gnosis is a spark lodged in the innermost depths of man, still an Idea at the end of an ecstatic albeit internal Transcendence, the contemplation of the Idea as of an ultimate atom of knowledge. Under an entirely other condition, that of the uni-lateral complementarity imposed on us by vectoriell thought, it would instead be a nonrepresentable knowledge, a knowledge that is nonecstatic through force of immanence, a messianic wave function acting upon the faithful, the ultimatum of faith that flows in the veins of Christ. Gnosis rediscovered the argument of the suture of science and Christ, a suture that can be opposed to all theology, against the Platonism from which it partly emerged. Christ is a generic knowledge—that is to say, one with no completeness or totality, a Law with neither substance nor logical form of messianity, but one that must be completed or fulfilled in relative occa-sional exteriority by the subject, by we-the-humans, as the flesh that must incarnate it and, moreover, the stance that must resume it.

Historical gnosis, at least qua Christianity become official, received the blood of Christ on the Cross, but did not welcome it as sufficient ontico-ontological substance, communicable to humans through a transubstan-tiation, like a process of ousiological metamorphosis. It welcomed it, as we have said, (1) as knowledge, not as substance, (2) as knowledge that is a constant or a quantum of messianity, a discrete flux that it was able to interpret as a Greek-style spark-flash. (3) And as generic, finally—that is to say, including its completion but not its closure in the faith of the faithful. For Christ can only bleed eternally with the blood of humans who are identified with the world or with God before becoming superposed with him. What permeates the body of the generic Christ is now the blood of the eternal suture that is messianity, the liquid element of the lived drawn from the subjects of the world. The blood of Christ is not that of a God offered up for sacrifice by religions. If it has been pressed, crushed, and tipped out, it is so as *to be infused* into humans, and it is their lived that flows out as the new generic blood.

THE SALVATION OF GOD

Obviously the good God of the gnostics, as we have long known, is not the God of the Old Testament, the jealous God who *admits that he can or could have* sacrificed his Son, that no other God could stay his hand. The good God under-goes as Messiah or equality in-the-last-instance of the Sons. We must nevertheless give the rigorous form of superposition to the gnostic suture and render it noncommutable with the world. One can imagine the sacrifice the old God had to consent to, in order to become messianic grace. The Other-God or the "Other-man" had to be able to superpose himself onto humans to transform them, to become principle of the Same, *idempotence—that is to say, fidelity through every trial*, to abandon his dreams of omnipotence and of the creation of the world. A forcing is needed, but an entirely immanent one. Against the philosophers of the universal God become generic, against the Jews, the Other-God relinquishes his hauteur, or, strictly speaking, refuses to send his Son to death. The Jewish God had to become immanent grace, the Other had to under-go as this Same that does not come back and for which his fidelity is enough. As generic, the Son is more than a dialectical "encounter," he is the superposition of man and the old God that will permit world-man to participate in messianity. In his person, God bends to the law of the necessary immanence of science, men abandon their individuality and under-go, liberated from their ego. What is salvation if not under-going within generic immanence, that is to say, the immanence of Christ? Not so much the loss of his ego as the loss of the sufficiency of that ego, the obedience to the New Law of messianity.

If it is God who is sacrificed, and not at all the human aspect, except insofar as it derives from this theology, then is Christ the revenge of humans upon a God that they have decided to reduce or sacrifice—an obviously religious and excessive hypothesis? The choice is between the sacrifice either of the Son or of the Father, but there is always an operation of the "reduction" of duality to the immanent Same. One historical modality of the sacrifice of the Son is the persecution of the hostage-Jew by the Most-High. From this point of view, Levinas repeats the Jewish model in very classical-modern, and very ancient, terms—for the Jew has always been the elect hostage of his God in living under his Law. As to the Logos-God, the equivalent of the Jewish hostage is simply the celebrated "subject" later to be nuanced as the subject of "Christian freedom." We oppose to these two figures the decline or the immanent abasement of the Father in and by the Son, and it is this that is undergone by the gnostics, who welcome the new knowledge as the Son or the mediate-without-mediation. The positive suspension of sacrifice is rather its placing under condition, not so much negative as underdetermining, in view of a messianic-oriented "Christianity." Gnostic knowledge is the knowledge of the Son, of equal Sons affirming their equality, that of Ultimate or Last Instances. For the Sons are certainly not content to sacrifice the Father as if they wished to take his place; they have changed place and times, they are the Orphans or the Ultimata and consequently the Prior-to-first. The Last Instance is not a substitute for archaic paternity, but its generic "fulfillment."

As for the Jews, are they capable of changing their infinite, and giving themselves over to the transfinite of the Messiah, to the unilateral or immanent opening? As with all conversions, with their dramatic weight, this would be a religion-fiction and not a christo-fiction. It would not be asked of them to admit an autoconstitution of the Same and thus an ethos of immanence, but to admit that God has for too long dissimulated through jealousy an ultimate property of thought, an "idemscient" property, an objective knowledge or gnosis that must be assumed by a lived in order to constitute a wave of messianity. It would be asked of them to choose between a God of servitude and a Christ of ultimatum. We are on gnostic terrain here.

God has changed states with Christ; he is borne by a new vector. Religions never emphasize enough that God and gods, like humans, must be saved, that "redemption" goes for them as much as for the humans who face them. If God himself has need of salvation, we can imagine the type of sacrifice to which the ancient Gods must consent in order to be accessible to messianic grace. The universal God of the philosophers must become generic, the Other-God of the Jews must sacrifice himself and not just remain in his heights, an entirely Christian god, strictly speaking, satisfied with sending his Son. The Other-God must be able to superpose himself with humans in order to transform them, to become faithful idempotence, to abandon his dreams of omnipotence. Through the act of faith the Jewish God becomes immanent grace, the Other under-goes like this Same that does not return. Salvation is not the repetition of creation, but just its suspended repetition, its christic resumption.

The frame of reference has changed. Substantial religions busy themselves with "triangulating" humans and their God, but Christ introduces another correlate, the world, and establishes a messianic axis, a unilateral complementarity, with it. Messianity is a vector or that transfinite throw for the world, not for Being or for a Logos better understood, not for a Torah better obeyed. God come into the world, his separate, Jewish transcendence reduced to christic immanence, it is the essence of salvation that changes. Christ accomplishes the "plan of salvation," which above all does not mean that he "realizes" or effectuates it, but that he submits it to its real condition, which underdetermines or "transforms" it. The old God was an organizer of hopes and a manager of expectations, in any case a redoubtable planner, which is the foundation upon which Churches establish themselves. What the separate and omnipotent God plans out like a transcendent or even transcendental captain of industry as salvation promised to humans, submitted to restrictive and selective conditions, Christ makes actual or active in its virtuality.

THE IMPOSSIBLE RETURN: TO REPEAT CHRIST THROUGH THE SCIENCE-IN-CHRIST

If traditional Christian faith is a belief founded on the resurrection and the re-turn of Christ, for us this repetition is highly problematic and must be clarified on the prior-to-first basis of messianity and its vectoriell ascending. Far from being a re-turn, a re-commencement, or a re-volution entire or in itself, it is instead articulated as and in terms of an imaginary quarter-turn, which explains the emergent origin but not the return, except as religious appearance. As superposition or interference of liveds that make up the "bloc" of unlocalizable messianity and faith, it is a question of the "complex" quarter-turns of a wave rather than one of the "thousand detours" from which Nietzsche composes the Eternal Return of the Same.

No doubt Christianity cannot do without a certain "return" that is combined with Judaic waiting and is designed to allow belief to test itself. The Law of monotheism in relation to which Christ is almost exclusively situated is not the only one, as we have said—there is also the Logos-Law of pure polytheism, which is consummated in the Eternal Return and which is necessary for a christic science. It would be wrong to underestimate the horizonal function of the constitution of this theologeme as pagan religion. Greeks, as the Apostles partly were (even Paul), cannot *not* think return rather than incessant under-going. Christ was understood as he who had come, but he was substantially understood as having to be reprised twice in order to assure the full salvation of the world. The Eternal Return of the Same is a question of memory or of that which comes back on a ground of forgetting, a return as forgetting or a forgetting as a turning. What could be more Greek and pagan, less Jewish, than the eternity of the actual present, the return of the Same as identical (or indeed as difference), the individual as proof and recollection of all experiences, of all divinities? The Christian return of Christ is a mixture between the unilateral coming of the Jewish messiah and his eternal return: the messiah must re-turn one last time. On the one hand, then, the messiah of the Jews would actually already have come once: this justifies Judaism for the Christian, and annuls it or renders it otiose. But it is necessary that he come back a second time: this justifies polytheist paganism and at the same time annuls it. The first coming of Christ is excessive in relation to Judaism, the second coming is excessive in relation to the eternal return of the messiah that the Greeks were able to imagine. Christianity as a religion is the synthetic, not quantum, midsite between the Jewish Law given to signify an infinite waiting, and the Greek law that is the program of a virtual return, through a forgetting of the "self" in its actuality.

The Christian Christ thus suffers for two different reasons: both the impatient hope for the second coming and the infinite waiting with which

Judaism impregnates it. It has been concluded from this that we must replace the failing or differing actuality of the Messiah with that substitute which is the Church or the actual community of believers. Because it very quickly tired of waiting for him, and became that monster of an instituted waiting, a calcified return of waiting, an *institutionalized messiah, a mystified body of believers.* The Church is built on a bad, too-historical understanding of Christ, upon a failure of theoretical analysis and rigorous thought. The Church has opportunistically filled this void, Dogma has taken the place of messianity and faith. Dogma is faith congealed into belief, closed up on itself like a world gone bad that diverts messianity into the exploitation of humans.

If the Messiah is in under-going, it is as superposition or continuation, or strictly speaking as "return" in the form of a Christ-thought, a gnostic faith or knowledge. The science-in-Christ is the only "return," the only "repetition" that is authorized for us, and it is a knowledge that resumes messianity as faith rather than creating a new history or a fantastic image. In the person of the Son superposing God and humans, a suspended superposition, the former bends to the law of the necessary immanence of science, and the latter abandon their sufficient individuality and under-go as generic. What is the real content of salvation if not the under-going of Christ within generic immanence rather than within the Church? Not so much the loss of his ego as that of the sufficiency of that ego through obedience to the New Law of messianity. If the pagan Law of return and that of waiting now give way to messianity as their superposition (and not their identification), another distribution of the advent succeeds its theological economy—what we have called the vectoriellity of messianity, or futurality. The idempotence of Christ is the constancy of an under-going, of an event that is resumed vectorielly, rather than being repeated, as the Same, despite the addition of a transcendent term that constrains it to become the Same. It is resumed without disjoining from itself or becoming multiple; it is the Same in a superpositional manner and not via inclusive disjunction. There are not two comings of the Messiah separated by infinite history, but only one, always the same incessantly and without making a thousand detours in order to come back to itself. He does not re-turn a second time (something that is also refused by Judaism) but comes actually (something that Judaism does not want, since it confines itself to waiting). He insists on under-going one-time-each-time.

The incessance of the Same is not the platitude of belief, but the simplicity of messianity.

Still, the "actuality" of the Messiah, actual in relation to philosophy and to faith as the belief proper to Christian mystics, is in itself virtual—namely, that the Messiah ceaselessly under-goes, this incessance being his actuality. Actuality signifies that the kerygmatic revelation takes place one-time-each-time but never ceases. The immanent messianic law is thus the law of nonreturn, the law of not being at home with oneself of humans forced to make do with the means that they have. The idea of a glorious return of Christ is prohibited by his noncommutativity with religions; it comes back once more to mixing the ancient and the new world, belief and faith—this would be to destroy the christic superstition of Logos and Torah, which would then coalesce as forces of repression. Noncommutativity is precisely the refusal of a double game, the certainty of the actuality of the Messiah for a science of religions that would not be a doxa adjoined to the religious imaginary, or the positive science of a spontaneously given object.

THE UNIVERSAL PHILOSOPHICAL CHURCH

We know that the Church was dogmatically constituted in its original struggle against gnosis; it forged the great majority of its weapons in the process of confronting the heresies that marked out its "development." It bears in its manner of thinking indelible traces of these repressive and conquering origins. As for gnosis, it is condemned by its very principles to an entirely other stance, a defensive stance, the acting of a nonact against the assaults of orthodoxies. But it possesses a freedom of theoretical means to defend itself and to deconstruct the appearances of ecclesio-centrism. We have demonstrated through its very exercise the theoretical right, the necessary axioms, to interpret Christ, *interpretation* being a discipline apart, not only in a science like quantum science but in gnosis qua power of invention. We seek the most fruitful and innovative interpretation, and here it is to invent Christ within gnostic faith. And it is a weapon in our hands, a weapon of defense or of resistance to the harassment of religions, all deadly to various degrees.

Many criteria may be taken into account for this thesis. One alone will · suffice for us but it includes all the others. The generic always goes not by way of any coupling whatsoever with a unity, but in binary manner from one unilateral complementarity to another, on the model of messianity and faith, from the subject-science to the faithful, from the Last Instance and from the theology that it underdetermines or brings down. This complementarity of the resurrected Christ and the faithful is the generic quantum of christic humanity. Measured by this yardstick, some ancient or recent solutions of theology can be brought back to their appearance. Gnosis has always had some very determined and impressive theoretical adversaries, but it judges that there are, to various degrees, interpretations at once inhuman and philosophical (the same thing) of the generic. They make the human real into a transcendent conception, because their theoretical means are those of the exacerbation of the omnipotence of God and the bringing down of man. Greek philosophy and Christian theology, it is almost the same combat, with nuances that permit one or the other to assure themselves of a community welded together by their conflicts of interest.

The extremely diverse theological interpretations are well known, and are grouped into several major groups, each with their own glosses. There is the authoritarian universalism of Paul, the classical Greco-Christian dialectic of the Church Fathers with its ecclesio-centric and Aristotelian becoming, and for us the most problematic divergence, that of scholarly theology. Also the way of dialectical christologies around and after Hegel. And finally the phenomenological zone, with the existentiell path of Heidegger, the path of called subjects (Marion), and finally the transcendental path of life and living beings (Michel Henry). This latter is a christo-centrism that brushes up against the generic human, but wastes no time in recuperating, without any true spirit of sacrifice, transcendence and the Trinity. Of course the latter is submitted to the immanence of the ego, but since this immanence is absolute and transcendental, it is not completely grasped as unilaterality or noncommutativity, and returns to an old-fashioned Trinity. These theological interpretations, all in the form of mélanges, are philosophical, neither quantum-theoretical nor (of course) particularly human. The Christ of faith cannot be the center of history but only the subtractive quarter of the Ascending, on pain of reestablishing implicitly the pagan circle of ecclesio-centrism.

In the conquest of dogmatic power, this was very quickly used against gnosis. Paul appears as the initiator of a falsification that was to open the way to the philosophical capture of the generic. It is he who is the founder or the "first" of Christianity—certainly not Christ, who is before-first, that is to say that he is the condition under which the foundation itself can be laid. He is the originary legislator of the Church in the forms of abstract universalism and egalitarianism—not Christ, who is at best his "Lord." Paul thinks the mélange of life and death, or at best their parallelism, whence the passage to the dialectic, whence his transcendent universalism.

It will obviously be asked how the construction of this christo-fiction is possible, or rather how it is real. The Resurrected of the Ascension is the mediate-without-mediation that produces the decline rather than the death of God. The lived does not sublate death immediately, as we might think, but uses it to abase God—it is generic, not universal; it is the Same and not the All. In the complementarity without the correlation life/death, Christ is the affirmation that two terms suffice, their unity not being a third term or their dialectic, but being rejected as evil or as world. The discourse of Christ is not that of life in general, it is the immanent messianic vector that contains death as its border or its Other repressing the excess of power.

The superposition of idempotence with the lived, the encounter of science and philosophy, of generic finitude and its sufficiency, these sutures justify themselves once they have taken place, but it has to have been done—it is a work, the work of faith. Quantum science on one side, Christ on the other, have realized it in an almost miraculous manner, even if the result legitimates itself only after the fact. Christ is given as Last Instance of human history or as messiah, but nothing justifies *analytically or synthetically* this construction that tears men messianically from themselves. Still, it must be seen whether idempotence does not prepare its superposition either with the human world or with the physical world, if as form or matrix of superposition, in any case, it is not capable of soaking up the lived, of forcing it, finally succeeding in this exploit. But how can a simple property of algebraic operations be sutured to transcendence? It is a work, a first work, already done, an initial "decoherence" that will only receive its meaning from prior-to-first messianity. This first suture is given along with philosophy itself as occasional but necessary material; it is the philosophical mélange of logic and of the real human lived or of the very world itself, which already virtually contains this suture (the latter is not just artificially imposed to justify gnosis). Evil is first, it is its own legitimacy; our task is different.

What is more, this suture also implies the harassing "deviations" against which we must rise up. The most constant is the conflation of Christ with a universal apostle of love, or even a theological "conceptual persona," the disastrous effect of his forced insertion into Christianity. Another, equally constant, is the recourse to Paul for reasons that have nothing to do with Christ. One might say that Christianity has not "flown," that this is a correct return of things, these old or recent recuperations of Paul and of his worst aspects by philosophers who make him the messiah of authoritarian militants, and found in principle the usurpation of Christ by the Church or by certain of its masked forms. This is an exacerbated invention of a second Christ by militants of a transcendent cause, of a truth that has lost its unilateral or lived force of the generic.

During all of this, Christianity, for its part, never ceases to "re-turn" as religion of mediation on the way to dispersion. It aspires to globalization because it is in essence or in principle co-constituted by the world-form, and is thus also less closed and vindictive than other monotheistic religions, being more formal than substantial. As institutionalized, as Church, it has received the transcendent side of the heritage (it has, precisely, made a heritage of it) without the power to transform it, contenting itself with the "development of dogma." As to ecumenicism, it is a laborious attempt to control the globalization, "evangelical" or otherwise, of the religious. But Christianity, like others, does not at all wish to liquidate its worldly heritage or its privileges, to pass them through the screen of the science of the idemscient faithful. It is incapable of delivering itself to the Law that, meanwhile, consigns it to an uncomfortable waiting, and to the Logos that delivers it to all the compromises with the world—it is on the Cross and has not yet risen or "ascended" to the side of Christ and with him. Whence this mentality of a religious heir, the need to assure oneself of a capital of ready resources. Not to forget that this twofold Greco-Judaic memory, encumbered with a thousand theological combinations, allows it to seal a new ideological alliance against Islam, which for its part demands only the subjection of an adversary. This is why we have need of gnosis, which *knows* only too well what the world is, and knows it through christic knowledge.

Thirteen

MESSIANITY AND FIDELITY
The Faithful of-the-Last-Instance

GENERIC INCARNATION, FUSION OF SCIENCE, AND THE SUBJECT

Christ is not a term in the conceptual architecture of a theology, but the name of a constant, the name of the messianic element of human life: the element of the immanence of the lived that finds its form in the algebraic imaginary of science, not in the imaginary of religions. It is the superposition, placing-in-stance, or first ultimation whose conditions and theoretical effects we have already posited (the suspended sacrifice of God, immanent or vectoriell fulfillment, the mediate-without-mediation). The christic constant has an effect of preemption over belief, and transforms it into a faithful lived. Thus is constituted a quantum of faith that is the christic constant incarnated in each of the faithful. Faith is belief transformed generically, neither singular nor total or global. It specifies the lived of any scientific subject whatever and makes the messianic vector into the vector of a transfinite faith that traverses beliefs without skirting around them.

The fusion of the quantum principles and the subject, that which we also call Christ as ultimatum, eschaton, or Last-Instance, is a phenomenon conceivable in a rigorous rather than miraculous manner or in the manner of a divine creation. It supposes an encounter not of the Jew and the Greek—that was valid before for the discursive material necessary for the operation of its cognizance—but of the idempotence of vectoriell origin and of subject-belief, with the christic generic redistributing or transforming unilaterally Greco-Judaic mixtures. The generic hybrid is far from being the first in history, initiating a new, or last, historical sequence—all of this is specular and circular. It is rather the prior-to-first event that reorients reverse-wise temporality, the arrow of time imposed by the world. Christ is the Last Instance, or, better still, the prior-to-first generic element as condition of every relation to the world. The world is the proper object of Christ that philosophy (Heidegger, for example) diverted and inserted into ontology. This event of generic fusion involves religions, obviously, but also the science that, *historically*, has not yet effectuated it in its mode, and that now serves as a model for the science of religions. For us it is a question of inventing, strictly speaking, a Christ who would cease to be imaginary in the religious, transcendent sense, who would not be an event in history (his anticipated or nostalgic repetition) but a new interpretation made possible by quantum theory. Our task, as faithful rather than as Christians, is to grasp, in the name of "Christ," this scientific or "idemscient" possibility, the real content of what is meant by his "imitation" as practiced by the mystics. We imitate a christic science rather than the historical Christ who serves as our "material."

Ultimation is the fusion, or the unified theory (something other than the conjoining or articulated junction of "opposites"), the constitution of the Last Instance as superposition. Or once again, in macroscopic and inexact terms, the identification of the subject with idempotence, the encounter of science and philosophy. Less than an operation, the Last Instance is an operated-on whose external agent is included in it. Christ gives himself, so it seems, as the Last Instance of human history or as the prior-to-first Messiah, and nothing justifies analytically, synthetically, or dialectically this emergence of a theory capable of tearing men away from the world.

We might examine the problem first from the side of philosophy, which, before being occasional but necessary material, is given as sufficient. Now, in the philosophical mélange of logic and the human lived, one can identify symptoms of this fusion as a simple suture in a transcendental form, a form imposed by an overseeing agent or a transcendent supersubject, a divine operator or theologian who ultimately conditions the transcendental and its false immanence. In this framework, ultimation appears as a simple redistribution of an All, but certainly not as a prior-to-first redistribution that would limit its authority radically.

The immanental fusion that creates the generic is prepared or "facilitated" in philosophy, but is not realized within it.

Another example of this insufficiency: philosophy supposes itself to be absolute, indeterminate, and virtual in its Idea, but actual and determinate in its particular doctrines. When it speaks of itself within its own particular doctrines—which it always does—it conceives itself both as entirely virtual and as an actual locality; it presents itself according to mélanges of these two characteristics. Now, a science of philosophy, or respectively of theology, supposes the latter in its All as well as in its particular systems, or even in its statements qua actual—or at least (to be more precise) qua actual under this *placing under immanent condition as mere macroscopic raw material.* Now, one cannot pass continuously from the absolute and its mixtures to the radical science of the absolute. There is a "leap" here, according to philosophy. The distinction is not between the All and the not-All, but between the virtual All in itself and the All under actual condition as material. What we require is the generic fusion of the Last Instance. The latter is actual but not effective or existent; it remains virtual in its actuality, it is an under-going. The redistribution as prior-to-first division is contained "virtually" in the actual All of philosophy, but in-One or vectorielly. We can simplify this by saying that there is an actuality of the under-going of the All, which undoes philosophers' oppositions between actuality and virtuality; there is an actual messianic under-going, but one that is not actual in the mode of existence. Just as for Christianity, which is actual under condition of Christ, but actual as an object of science might be, not virtual as for the theologians, who move within a possible or imaginary Christianity.

THE UNILATERAL COMPLEMENTARITY OF SUBJECTS

Between science as Christ and philosophy, between the flux of messianity and the individual that it draws-under to the hell of the world and transforms into one of the faithful, the "subject" is an ambiguous notion that has many aspects or passes through various phases. Let us recapitulate.

On the one hand, there is the prior-to-first Christ as generic subject-science or Last Instance, as superposition of the minimal algebraic condition, idempotence, and the lived that it neutralizes. This lived is of an indifferent doxic origin, but becomes generic or nonegological, nonindividual; it has passed into the state of immanent messianity. Its generic form excludes the consciousness-form or the ego-form, which are rejected as worldly representations, or representations contrary to the wavelike nature of messianity. There is thus a generic lived of Christ as science: Christ is not a singular individual of the world nor an ontological entity; his universality which takes humans in charge is of an other nature than that of philosophy, which "takes into care beings as a whole." But it can always be conflated, by philosophy itself, with the transcendent subject of philosophy. Philosophy has every interest in conflating the two states, messianic and philosophical, generic and general, in the name of the "subject," and hence identifying it with man in general, or abstract humanity.

On the other hand, at the other extremity, that of the world, there is precisely the worldly philosophical subject that believes in itself or would be "sufficient." But its function and its usage are now complex. To philo-theo-logy and their diverse subjects, the science-in-Christ accedes in principle, removing all sufficiency, making every transcendence fall into-immanence or into-messianity. It also "searcheth the reins and the hearts" of the worldly subject, but searches them so as to transform them, less to "con-vert" them than to "uni-vert" them, to turn them toward the generic or toward the immanence of messianity. Thus the philosophical subject is given to this science-in-Christ as symptom or as already the object of its particular action. We also call "occasional" this free variable of the messianic function that plays the role of a symptom.

But between these two "subjects," one as fully generic Christ, or messianity, the other as simple occasion of generic salvation, there is the faithful subject configured by messianity, the messiah that haunts the world. This subject is what is transformed into a symptom by messianic immanence. It is the faithful subject or the subject-in-struggle against the beliefs of the world. Now, it is principally this subject-in-struggle, this faithful grappling with the world, that bears within it all the ambiguity of a double interpretation. This subject is the one that is saved *and therefore* can once more be caught up in the world and its attraction. Salvation cannot be an absolute or "religious" act, without premises, as creation ex nihilo claims to be. It is indeed a grace, but we distinguish between an absolute, authoritarian grace fabricated with Greco-Judaic presuppositions and a radical grace that takes humans at their vectoriell root. How could we be "saved" if we are not already saved in knowledge, without being this knowledge of a salvation that does not yet actually know itself? If messianity boasts a certain action, it is that of saving humans and making messiahs of them. We do not conceive a divine and individual messiah coming into the world, or, a little differently, a historical Christ making his return at the end of history or of time. To be damned is to *believe* that nothing can save us because we do not know how, by what means and through whom, we are saved. We are the test of a salvation of which we know almost nothing. However, we have the feeling of having been saved despite our long-maintained ignorance of the means necessary to have been saved. "God made man" or "man made messiah"?

THE OBSERVER SUBJECT

The faithful are not faithful immediately; they are often agents guided by religious or worldly concerns. What about the soldiers who put Christ on the cross—will they be saved, and on what condition? There is an objectivity of the observer of the christic science that remains what it is macroscopically, but also an inclusion in-immanence in a unique messianic wave function of this observer become his faithful clone. As a function of this doubling of the observer of the christic science, what should we call this "individual" subjectivity or faith, if not "messiah" or "messiah-existing-subject"? A science, be it christic or "non-Christian," must in any case eliminate the philosophical foundation-subject, must be a true science that recognizes its contingency, a procedure whose objectivity cannot justify itself through a program of autofoundation. And at the same time, for a twofold reason that makes of it a quantum science of human (for example, religious) stances, it must include in a unique state vector a subjective complement to the procedures.

Christ is indeed a generic "subject." It is not because here he is not an (involuntary) founder of religion or a historical figure that he is an anonymous symbol. Not only is science a work of humans rather than of gods, and one that merits "subjects," but if we call generic stance a procedure operating as the underdetermining condition of the scientificity of a discipline, but a procedure that has a complementary subjective aspect, only an aspect, we give a rigorous meaning to Lacan's "subject of science."

HOW THE FAITHFUL ARE CLONES

As Last Instance, messianity is a superposition and thus a process of vectoriell resumption. The reprise is a part of its very essence. Each resumption one time each time underdetermines the theological double transcendence that it traverses, and from which it is relatively inseparable, abasing its power to the particulate state. When the transcendence concerned is that of the ego or indeed of consciousness (even so-called transcendental consciousness, because of its general context it contains double transcendence), since it is inseparable from such a corpuscle as Being is from beings, the particle obtained or included in immanence is that which comes from the philosophical-type subject. It can still be called "faithful-existing-subject" or simply "faithful." This cloned faithful that phenomenology would have called noematic and that quantum theory calls a particulate clone has its essence or its form in faith, not at all in belief, which is but an occasional material for faith. Faith is thus faith of belief—for example (but not only), religious belief—underdetermined by messianity and its resumption.

Now, how is this underdetermination as particulate of a transcendence of corpuscular origin possible, this obtaining of a particle through bringing down or simplification of its transcendence, now included in the immanence of messianity? We know that messianity is a wavelike phenomenon that traverses immanently, in a "tunneling" manner, and thus by *ascending, but without transcending a second time or once more*, the infinite mountains of transcendence, not in the way that demons or sheep who have escaped their enclosure might have done. We shall call tunneling-effect, in general, the action-without-transcendence or the vectoriell ascending of immanence across transcendence. But the effect of messianity is not summed up in this passage alone, this passage of vectoriell immanence across transcendence detaches from the latter a simplified form that does not add itself to transcendence; it is instead its transformation through subtraction or through the bringing down of its doublet nature. Messianity is wavelike and thus belongs to vectoriellity and to the imaginary number denoted by $\sqrt{-1}$, which has a subtractive effect. The effect of this immanent passage is to reduce the corpuscle to a simple transcendence, to suppress the original doublet. We shall say that simple transcendence is the clone-form of double transcendence, that the faithful-subject is the clone of the transcendent(al) subject of the philosopher. Cloning is a *repetition, itself immanent, of transcendence, and is thus subtractive or depotentializing.* Rather than being spontaneously overpotentialized as philosophical repetition is, it "repeats" its material givens at an apparently lesser power, as simple and weak. It is a question of a cloning, and six differences can be remarked between this and a wordly being.

1. A material must be provided. Cloning is not at all a creation, and destroys the fantastic concept of creation ex nihilo; it is a transformation obtained through unilateral and generic conjugation of the variables of a material.

2. The entity obtained, from the macroscopic point of view of the material, seems to reproduce the original identically, but in a weakened form, without possessing all the properties of the original. In reality this opinion stems from an insufficiency in the analysis of the original situation, which is always a double transcendence that has been massively simplified in a homogeneous manner, like "Philosophy" itself, and from a conflation between a simplicity that really originates in complexity and a "doxic" simplicity of philosophical origin.

3. The resemblance to the original is, moreover, ambiguous to say the least, and ceaselessly fools philosophy and theology. The clone is unifacial from the point of view of its messianic immanence; it is also foreign (clones are the Strangers among us), but it is given as bifacial, and thus as resembling a being of the world, a resemblance that is an objective appearance since the clone and its original material are only half or partially of the same nature.

4. It is a question of a scientifically conditioned operation upon philosophy. An essentially quantum superposition is necessary in immanence to obtain this effect in transcendence.

5. The true force of the clone is that it detaches itself from the givens of the world, to acquire a more developed theoretical and ontological status: it is onto-vectoriell. The faithful-subject is under messianic condition and can transform the world, these are its "works," because it is the first to be transformed by messianity.

6. The faithful is "in the image" of Christ and not at all that of God. Cloning is the passage of subjects from the image of God to that of Christ. This may also be called salvation through fidelity and its works. It is obvious that this overthrows the foundations of the Christian imaginary, and that it must be reoriented toward a christo-fiction.

THE MESSIANIC UNDER-GOING, ITS ACTUALIZATION BY THE FAITHFUL

The question of the new lived, the faithful lived, contributed by Christ to humanity, is fundamental in changing practically our relation to the world. It signifies that the question is not: what ought I to expect of the coming or the return of Christ as of a providentially sent angel, how should I await it and when will it come, what can I hope for, me, a worldly man? These are questions of belief, philosophical and desperate questions, and *any response, and even the questions themselves, here spell the immediate destruction of messianity*, because the means to think the return, and the response in return, immediately contradict the way in which faith constitutes itself or acts as messianity. Its generic-being and its noncommutability with philo-theo-logy can only "exist" virtually, outside the worldly sphere, in the radical immanence that does not "ex-ist" in the open of the world but underexists or is affected by a secret.

However, it is still possible to formulate oraxioms concerning the nonbeing of messianity in which we participate, its inexistence, and its secret, since the secret can be oraxiomatically formulated even though it is foreclosed to theology. This is why we say that it under-goes as generic rather than be-coming it philosophically, rather than over-coming like a hero or a "savior," and also that it does not intervene like "providence" in the course of the world. Nor us—for in reality it is a question of us, we "Christians": we do not have to *become* or *be* in the substantial or indeed subjective sense of these philosophemes, as one is or becomes a philosopher. We have to under-go as what we are, faithful who can sometimes be "without-religion." If there is a law that messianity does not follow, being unlocalizable, neither local nor global, it is that of the course of the world or of history. Given its actuality, which is not historico-worldly but entirely an actuality of immanence, and despite its underexistence, subjects drawn from the world cannot say what happens in the mind of a Father who would send them, or in those of the sent—all the "whys" that would motivate an anthropomorphic finality—but only how he acts, that is, how he refrains from acting or existing, the better to act through this immanent retention, this underexistence of the vector in which he is held and which suffices to act on the faithful that it extracts from the world. It is a problem to resolve through its own transformation by each of the *faithful*, not a philosophical question addressed universally to individuals, or an existential one.

Through its messianity, the science-in-Christ ceaselessly under-goes with the incessance of its coming: this is its form of actuality and eternity. There is no "return" of Christ, still less an eternal return, but an eternity of under-going that suffices, that is not an empty and passionate Judaic waiting. Christianity certainly is effective or worldly as an imaginary formation, as is Christian belief, like every object of science; but Christ and salvation are actual despite their inexistence or their worldly virtuality. The Messiah is not an object of belief or of an infinite waiting. The entirely internal and spiritual waiting of the Christian for a Christ who would already have come, as *a messiah of memory and thus "for memory,"* is traced and sublimated from the Judaic duty of waiting for an inactual or impossible messiah, a terrestrial savior—but simply reformulated according to the Greek Law of the Eternal Return of the Same, and thus doubly a mytheme. A Judeo-Greek Christ is not an occasion inserted into the general fabric of a history that it would be content to tear up, for to tear up a continuum is not to determine it and suppose its repetition. We do not conflate the repetition of historical or subjective occasions and the reactivation of faith to which they give rise—these are two unilaterally incompatible but complementary essences.

HOW THE FAITHFUL FULFILL GENERIC MESSIANITY WITHOUT BRINGING IT TO A CLOSE

Must one be a believer to practice this science? "Believer" cannot mean "faithful." Believers, we are in every way, and always too much. No "personal" or "existential involvement" is necessary in real or generic conditions. Personal involvement—and there can always be some—takes place only under the occasions of the establishment of this science. The installation of a subject into the stance of the two quantum principles, even when motivated by a vague "scientific interest" for Christianity, permits the assumption, in the order of knowledge—that is to say, the axiomatic of the Christ-science—of the basis of the lived generic constant, the messianic vector.

But one must complete or accomplish this prior-to-first decision and make of it a constant, and the rest will come of itself, or will take place, for the essential, according to its mode of necessity. This completion of the subject-science as constant signifies that it vectoriellizes and unilateralizes as messiah the transcendent subject that is proper to this domain of objects, and that proves to be existentially involved but without this being determinant. Prior-to-priority, some lived, wandering the world, must be carried to idempotence as immanent messianity and to the noncommutativity that make it generic subject, as faithful, amorous, dominant, and so on lived. The generic sciences have need of a lived that is each time specific; politics needs a proletariat or an acting subject, erotics needs an amorous subject but cannot be reduced to it, aesthetics needs an artist subject—all subjects philosophically neutralized so as to become generic. It is thus not necessary to be philosophers in order to establish in principle if not in fact a science of philosophy, except obviously for the knowledge of technical details. And hermeneutic all-faith is not necessary to explain the religions or the Gospels and to reduce them to a constant. These generic subjects, and above all Christ, we must conceive them as quanta of lived algebra—that is to say, quanta of the force of fidelity.

As soon as he is regrasped and statuified by the religious, Christ as emergence of this science of salvation seems to leave his work in some way incomplete, as something that has been able to arouse the hope or the imagination of a return, of a second, no longer militant but triumphant attempt. But if the faithful are necessary in order to make intelligible this knowledge that they are, the faithful are also necessary insofar as they complete through their fidelity this science-in-Christ without bringing it to a close, insofar as they sum these vectors without exhausting it. The generic duality of Christ, this immanent and rigorous messianity that acts on the faithful without passing via the world, if it is badly interpreted or grasped externally from the point of view of philosophy, simulates a return or a doubling in the mirror of Christ and of messiahs as individuals extended to the dimensions of world-history. As if there could be a doublet of Christ, or as if he were a doublet of himself. The famous "plan of God" comes from the myth of return and its worldly and apocalyptic horrors. What Christ teaches is not a hope of return, to put an end to an unfinished task, a creation that remains "broken down," a first unhappy effort, as the divine creation of the ancient gnostics may have suggested. There will be no return, nor even a new apocalyptic turn of History. Messianity understood rigorously, with the means of an order more scientific than mythological, exhausts its essence one time each time, and only multiplies its effects as a function of occasions. However, the science-in-Christ necessitates that we no longer go into the world other than through it, that is to say, through faith; and its strength is entirely that of a straight and unifrontal momentum. In view of making oneself a generic subject—but this is not a transcendent project or a program of Providence—messianity, being prior-to-primary, is necessarily "completed" by the faithful under the law of immanence that acts on them, the faithful that it subtracts from the world, without mechanically or transcendentally constraining them. The messianity of-last-instance underdetermines the faith of the faithful, extends itself into the world and beyond the effectiveness of the messianic under-going. We the humans, we are the unique face of messianity, we are unifacial messiahs and not individual messiahs, founders of sects. It is up to we the faithful to complete or actualize messianity; we are the faithful of-last-instance because we are acted upon, no doubt, but by the christic nonact.

Christ is thus not the immanence-all of certain mystics, but a generic immanence that requires the lived of faith to under-go as messianity. Far from being an external and transcendent coming, this under-going presents itself, however, as an exteriority to the philosophical and theological lie, and must find means for the struggle. As new Law, it is now not split or doubled in the mirror, sometimes along with the Word, but unilaterally divided into two complementary moments. Not a doublet but a unilateral complementarity. Each believing subject in the world or in their religion must now fulfill in their person, or assume, messianity, and make themselves faithful-in-struggle against the harassment of the world. In other words, only vectoriell immanence is not completed or subjectivated, not yet lived as faith transforming the world or "moving mountains." The faithful is himself a messiah on the way of fulfillment, who assumes the world only to deprive it of its sufficiency; he fulfills the new Law of idempotence or completes it by adding to it a subjective complement that relays messianity.

APOPHASIS AND CATAPHASIS

The Judaic messiah and the Heraclitean flash are two great founding myths of the West that have only rare points in common, generalities rather, centered, as we have seen, on mediation. They are always awaited and received or receivable, but in such a way that there is a certain circle between the awaited and the received/that which is come, a certain differed reversibility through the infinite and ultimately unreceivable waiting for the messiah, or else repeated by that of the philosophical flash already come back. But their addition as the states "Torah" and "Logos" is pertinent for the vectoriell fulfillment of Christ. Their superposition in the generic Christ yields a unique *irrecusable and irreceivable* state vector, *cognizance of which we seek.* The generic Messiah or humans in-body are not awaited, do not make a circle with anything of the world, and yet under-go with a certain degree of indetermination, freedom, and "unexpectedness" or ignorance, rather than their simply coming back in parousia.

The addition via superposition and, moreover, noncommutativity of the Last Instance creates the inevitable or "objective" appearance of a vectoriell subtraction (an "apophasis") or of an underexistence of messianity in relation to philosophical and theological control. But the prior-to-priority of messianity, perhaps unlike that of the subject-in-struggle or the faithful subject, is not originally a contingent decision that could be interpreted dialectically in two opposite ways. Messianity is a form that is a priori through ultimation or the decided-without-decision in which all of our acts, whether scientific or faithful, are immersed. The retreat of messianity from and in relation to scientific positivity and philosophical spontaneity is an apophatic appearance that is itself philosophical. The positive essence of subtraction is rather, in its origin, an under-going that "ontologically" brings down philosophy or theology to the state of clones or particles, pushes back their priority, even though they in fact do not at all abandon this priority, conserving it in the form of an appearance. Cataphatic and apophatic, this classical duality of mystical theology that developed in the unique milieu of double transcendence is now understood through an entirely other context than that which it is enveloped in, and whose aspect is at first sight brutal and physical—the wavelike and the particulate, both terms that display a certain crossing of theological opposites. The duality that is substituted for that of philosophy overthrows the Christian logic of the rising and the descent of the discourse on God or indeed Grace. Vectoriellity is a christic and not a theo-christic device; it is formed of the simple ascending of vectors superposing themselves on others, but this creates a bringing down of double transcendence or of every possible

Logos, to which the phase tends or which invites it, but which is arrested and fallen-into-immanence.

The "step-backward" is now that which Christ forces theology to take despite itself, a bringing down. This is imposed upon it by the faith that under-goes to meet it. Judaism awaits the messiah, and defines his coming in terms of this waiting, but it is now the waiting that is a symptom or an effect of the under-going. This ultimation has the immediate effect of the uni-version of the arrow of time, the effect of messianity. One might see in this, of course, an apparent withdrawal of messianity; but in reality it is messianity, the futural, that slows down and above all weakens philosophy as present and past. We understand the vectoriell ascension of Christ, his insurrection, as the true apophasis in its positivity. It commands the descent of the Logos that will henceforth be in decline, the bringing down of power. There is no essential withdrawal for Christ himself: messianity has a positive subtractive-being, it is insurrectional by nature, it is representation that retires itself or is rejected under the effect of the messianic under-going. But the prior-to-priority of the insurrection is accompanied, at the other extreme of the process of appearance, with being thought according to the limit or the impediment of representation that would condition it, whereas the forced retreat of the latter belongs to it qua appearance of real movement.

THE FAITHFUL OF-LAST-INSTANCE

Christ as any faithful whatsoever? This is not what we are saying. Neither subject nor object of this science, neither individual à la Kierkegaard nor believer nor member of the Church, who is Christ and each of the faithful following him? What transmits Christ to us is the figure of a faithful equal to others, a "common" but generic being, for each of us, just not any faithful whatsoever in a flat, banalized sense. The Church is too often the assembly of any faithful whatsoever under the norm, hierarchized and organized according to a law of inequality in relation to God, who assures the "mediating" and separating function of the priest. Are the faithful in-Christ exemplary or samples of a theological type? Certainly not. Christ is a faithful of a generic nature, not in the sense of a community or generality. He is a genericity, not a transcendent one but a genericity via immanence. His universal coming for everyman is not that of an All, even one rendered "open" through philosophy (Bergson, Heidegger, Deleuze), nor that of an individual with a singular destiny, a knight or a hero of faith (Kierkegaard)—all rebellious but ultimately Christian solutions.

The generic is a way of sterilizing and contracting, through their superposition, the dualities or doublets of philosophy, without reproducing or replicating them. How does one contract an indifferent philosophical or theological duality like that of God and man, like the theo-christo-logical doublet, if not by contracting it punctually, as in the death of God or the death of the subject? The withdrawal of God or his death is a surreptitious doubling by invagination of ecstatic transcendence. Christ is the superposition of God and the subject; his immanent work consists in "adding" them idempotently in his person, so that their distinction, their discernment, and their individuation become impossible. Superposition is obviously not a total and imaginary identification of God and the subject, each mirrored in the other—an identification that is at the bottom of the mysticism and so-called German idealism in philosophy. It proceeds otherwise, without specularity or speculation. It reduces the first to a state of vectoriell flux of vanishing immanence, disappearing from the horizon of transcendence, and the second to the state of a subject acted upon by this flux. In Christ, God and the subject have the same duel Lived, one sole lived-for-two that they share without dividing it. Whence a contraction that can seem like a subtraction or a kabbalistic withdrawal of God into his transcendence. As to the second term, messianity reduces it without denying it, including its transformed transcendence in its flux. In Christ the duality of God and subject is constituted in a unilateral or complementary duality densifying itself in itself in the form of a unique vector of immanence for the first and, for the second, as messiah-subject stripped of its egological form and of the predicates that accompany it. The lived has lost its ego and the ego itself is destined to interfere with itself.

Why is this equality of Christ and the faithful in messianity or in faith not a general common-being? The principle of noncommutativity obliges us to distinguish and to name Christ in his science otherwise than as "simply" faithful. He is the faithful of-last-instance, or the prior-to-first faithful. And every faithful, if he follows Christ, on the one hand, is no longer any faithful whatsoever, but feels himself to be community and generic body—this is the stance of faith liberated from the hierarchy consubstantial with ecclesio-centrism, from churches and sects—and on the other hand he himself also deserves the name of faithful of-the-last-instance. To say it in yet another way, Christ and the faithful are an ultimatum posed as the ultimation of faith or as the faithful-in-struggle. An ultimatum that is the content of the new kerygma as an announcement made to the world of a "sterilization" through fidelity of the great spiritual cities, the stubborn Jerusalem, the transcendental Athens, the mystical Byzantium, and the Catholic Rome that reunites them. The eschatological faithful is the messiah, and any faithful speech is an ultimatum nailed to the doors of the Churches.

Fourteen

FAITH HARASSED BY BELIEF

BELIEF AS DECOHERENCE OF FAITH

What, then, distinguishes the individual subject determined in-the-last-instance as faithful from the old subject of philosophical belief from which he is extracted, as its clone; what distinguishes the kernel of the faithful lived from its metaphysical placenta?

1. Its deindividualization or generic nonegoism, which it owes to messianity. A subjective lived necessarily human in origin but become an objective procedure is not denied or reduced to an object, it has just lost its egological form, its form as transcendental consciousness. Science here is the science of living beings or of believers, but the science of living beings is not itself living—this would be a vicious circle—rather the science of "spiritual" life is itself lived-without-life. Since the living being is the circle or the fold of the lived and life, the christic science of living beings knows nothing at all of life, it requires it as a lived that is unknotted from life, with the latter remaining sutured to the lived. Life gives rise to philosophy and theology (Michel Henry) but our problem is still elsewhere: it is not to found a new theology, whatever one might think, but to render scientifically intelligible this premature and precipitate "science" that is theology. A non-Christian science has to be not believing or faithful without further specification, but generically faithful.

2. Its nature as christic lived. We already know the two procedures (superposition and noncommutativity) that make it a science on the quantum model, but what is it that specifies it as non-Christian? Its essence of faith. Obviously it is more generally a lived, but this lived is determined according to the sayings (*paroles*) of Christ as path or way, as memory of a futural coming, as fidelity. As messianic superposition, faith is the essence of every fidelity, and one of the specific traits of the generic human lived, which has a plurality of them. Each generically human science requires a trait that specifies the subject concerned in the form of a lived of a univocally faithful nature, whether it is the amorous lived, the lived of the artist, the lived of the political subject. Faith in the generic sense has almost nothing to do with belief, which is its transcendent and worldly derailing; it is not faith in . . . a person, an idea, or an event, but immanent faith. Just as lived and subjective faith does not refer to a subject but to a subject-science, it does not refer to an object, even an interiorized and idealized one, it is of the nature of immanence, which is neither an all nor a singularity but a superposition. *Vectoriell superposition is messianity as fidelity of immanence-(to)-itself.* The Christ-event makes a new science emerge because it is the science of the world as object of belief, a world of which Being, the One, the Idea, or *physis* are modalities. The world is the gnostic object par excellence, globalization or not. The conflation of faith and belief, of messianity and the world, is the disaster that affects faith as revealed by Christ, the Christian confessions being the agents and the consumers that render "decoherent" faith as belief. Quite obviously this destructive decoherence is the real content of "original sin," which is its mythical projection.

THE QUANTUM INTERPRETATION OF ORIGINAL SIN

Can "original sin" still have a meaning or receive a generic or quantum interpretation? If there is an "original sin" that is the fault of generic humanity, its fallibility or its temptation, it resides in the extreme ease *in principle* with which faith can be destroyed by belief, belief in the world in itself and in its relevance to the things of faith. Faith is messianity, it is lost or destroyed in belief and its temptation, which is its confusion or macroscopic identification with the world, not even its becoming-macroscopic, rather its destruction. The exercise of faith as decision fallen into its own immanence is arrested or obstructed by not only that which should only have been the occasion of its resumption, but that which has become the origin and foundation of its thought and its life, the world can only destroy it in its principles. At best we have life and movement in the milieu of the world whose most general form is spontaneous theology; we do not have the lived of the vectoriell reaction that "moves" mountains. For faith itself, mountains, in their immobility, are unlocalizable. The mechanism of belief is precisely the vainglory of religions, the loss of the generic through an excess of transcendence, which makes them interpret sin as a fall, when it is more an attempt at elevation and overgrowth, a negation of the generic state of humans denying the conditions of their salvation.

To speak of the evental encounter of quantum theory and the subject that produces generic humanity, each religious Law has its intuition and its interpretation. The Greeks, since Heraclitus at least, have the flash, the source of all philosophical forms of light that give and withdraw, shine out and fade, it is the aesthetic intuition and ontological seduction. The Jews have another experience of the real, a "Sinaic" experience through the voice and through an affect of the absolute Other, the ethical intuition of an absolute transcendence. Voice and light are intuitions that hesitate between metaphor and concept, but for us they are only symptoms, for they obviously have not taken on vectoriell form; they are prequantum and lacking in "formalism." They are limited or inhibited by subjects as spontaneous philosophers, whose transcendence resists the traversal of waves of voice and of light, and whose egological properties transmit messianity poorly. These properties, which religions interpret as egoism of the flesh and sin, refuse the passage of messianity and destroy it. The logic of our interpretations leads us to say that the rigidity of the sinner is his original "macroscopic," that is, worldly, nature, and that original sin is understood as destruction of idempotence and superposition, as decoherence of fidelity, destruction of the very conditions of messianity. From the point of view of philosophy and religion, every wavelike or vectoriell phenomenon is perceived in terms of the world, under conditions of punctuality, circularity, or globality. But messianity is only an event if it has the continuity of phases and does not disappear punctually in itself or into a subjective interiority—it must under-go like a flash that is not extinguished in itself like a light that withdraws into itself; it under-goes as the secret of a "half-light" that calls for a theoretical explanation rather than the doubling operation of a clarification, an elucidation, or perhaps the simplicity of a clearing. As to the voice, if ultimately it does participate effectively in messianity, in so doing it would cease to affect itself as voice of Being, but also to signify through its infinite reserve and its transcendence. It undergoes as voice "in-phase" and without emphasis, underheard.

MAN MADE MESSIAH AND THE STRUGGLE AGAINST BELIEF

The vector of messianity is a virtual and therefore incomplete gnosis, a science that must incarnate itself in human materiely or dress itself in faith, a knowledge called to become a neutralized and generic lived, a subject-without-ego. Of this lived obtained via superposition, it will be said that it is cloned from the lived of the philosophical subject or pre-empted over it by the vector, rather than the vector being schematized in it; but as idempotent, it no longer owes anything to the form of this occasion. It is enough that the philosophical occasion (that is to say, we, the believing subjects of the Church or of the world), that we should spontaneously place ourselves in the position of sufficiency, for from and "in" this immanence has already emerged a messiah-existing-subject, this time in a stance of struggle against the world and belief. We do not re-act against the world—it is the world that resists the transformation imposed on it. The generic immanent lived under-goes messianically the transcendence of the world, unilateralizes it, or makes it fall into vectoriellity. If transcendence were really and absolutely in itself, to that extent we could believe that the Messiah under-goes as transcendental and as schematizable in the human as anthropological given. But the process in the matrix is immanental and not transcendental, and the transcendental version of messianity inverts the real process (which is not the inversion of the apparent process). Thus the simple vector of messianity must be fulfilled or completed by the faith of the faithful-subject, by the messiah in his struggle against the world. Becoming messiah or faithful in-the-last-instance implies or conditions the struggle against "Catholic," or more generally religious, paganism, against the belief and fetishism that objectivate faith transcendently. The old Greek and polytheistic paganism was a paganism in the eyes of monotheism, but it is now the hypertranscendent monotheisms that are vainglorious substances in relation to the faith that transits Christ. It is a matter of incarnating subjectively in we who are also the occasions of infidelity, the Christ who is a matrix (not a center) of all incarnation. To be exact, it is less an imitation in rivalry with a Christ supposedly already complete than his unilateral "imitation." The content of Christ is that superposition with messianity that we, subjects of the world, are forced to accept as a possibility of failure. The imitation of Christ is not that of a Master or a Hero; it is done, it is "made," through that superposition of messianity with belief thus transformed, and as a cloning of a faithful-subject in struggle that derives from it. It is the happiness and the suffering of being torn from the world qua sufficient, not the eternal return into the self or into the ego of pious interiority. The messianic flux of faith must be completed by our becoming-faithful in a state of struggle against that which we are always too much: believers-in-the-world. If God is "made" man, couldn't we say that we the humans are "made" messiahs, against the times, against all evidence?

THE THREAD OF FAITH AND TEMPTATION

We know that the structure of unilateral duality is susceptible to a double interpretation of the subject that it contains as noema or clone fallen into messianic immanence: it is interpreted either as subject that knows itself faithful or determinate in-the-last-instance, or as objective appearance of a subject that lives as in-itself and as delivered to the world as a function of its attraction = X, which is the force of evil. As soon as generic messianity "surpasses itself" as subject of individual origin and configures a faithful subject, it is inseparable from this double interpretation. It is in this that Christ cuts out messianity "by the sword" from the human subject that is, nevertheless, indivisible, but solicited by a double interpretation: philosophically believing or generically faithful. It is not the One that is divided or decided, only to be finally reconstituted. It is in fact the One in its idempotence that divides and imperceptibly detaches the saved subject from the subject that has given its belief to the world under the constant or structuring attraction of original sin—that is to say, the dominant force = X opposed to messianity. Faith is obviously a duality with only one face, but one that must, in principle, become incessantly double-faced and fall into ambiguity. To perceive generic messianity and the world-evil = X as a duality, with the messiah "between them"—this is precisely the illusion of theology, which confuses the messiah with a messenger, the mediate-without-mediation with a mediation "between" opposite terms. On the other hand, the state of fidelity under-goes when the faithful subject knows himself to be detached, through messianity, from the unfaithful subject, now seeing in the latter his worldly double or his caricature, without placing his fidelity in it. Unilateral duality is not only that which passes from immanence to transcendence; it is the fact that this duality passes or is concentrated *in* the cloned faithful subject, which can also always be the sufficiency of the in-itself. How can the "creatures of God" or "children of the world" that we are not be born a very first time in faith? If one has not forgotten that messianity makes itself unifrontal or unifacial, one will admit that the edge of the sword is that of faith or, better still, of messianity. Strictly speaking, messianity as radical immanence cuts into theological double transcendence, but theological illusion and philosophical error are already in this interpretation, which transfers the duality of that which is decided, the theo-christic doublet, onto that which decides and which would be of the same model, the biface of a sword or a weapon. Messianity does not act: this is why it cuts unilaterally. Messianity depotentializes the strong force, simplifies it; this is the only way in which the weak force of immanence can cut the Gordian knot of double transcendence into which the world is twisted. The sayings of Christ have been taken and understood as the dogmas or sayings of a prophet, as given or as containing, from the start and in themselves, the meaning that they must have definitively. Not only is there a religious and mythological fetishism of these logia, but theology, obsessed by philosophy, has neglected the fact that knowledge of Christ is not at all complete and given, ready-made to enter into a system—it is a process of cognizance in progress. If these sayings do not announce a plan of salvation, but rather describe it, at least ideally, it is indeed because there is a becoming or a process in the kerygma, and because it is up to us to fulfill it, to verify it in our acts and in our lives. These are not commandments or norms, imperatives, but variables for a future life, coordinates to define the lived of the faithful. Christ's sayings are the givens of the problem, not its solution; as for the solution, it is concentrated in the becoming of the faithful.

There are three ages of Christ: the prophets, the apostles, the faithful. From the religious point of view, Christ cuts orthogonally the course of history and redistributes its phases otherwise: the age of prophets (the preparation), the age of apostles (the announcement), the age of the faithful (the implementation). Each of these ages has its own way of being contemporary with Christ: the Church is founded upon a contemporaneity with his historical past, it is the reified preparation; the present is that of the apostles or of theologians with their different testimonies; and the faithful are the secret apostles of Christ, contemporaries of his promise. The faithful successor to Christ qua promise is not the apostle or the disciple, he is contemporary with the promise and with futurality, but not with the present like the apostles, nor with the past or with transcendence like the prophets. In this gallery of figures, the philosopher and the theologian are hybridized characters, haunting the world and condemned to make a tradition of their wandering.

BEYOND ATHENS AND JERUSALEM—AND ROME

1.   Under the name of Christ, it is a matter *at once but in-the-last-instance* of the real of faith or of faith as real object, and of a cognizance that is itself faithful of this faith, or of the faith-in-person of the new faithful. The generic lived or "Christ," cognizance of which we seek in the name of faith, is itself a "function" of two variables. For *the Jewish and Greek givens are the variables or properties of Christ to be conjugated under Christ as "imaginary number,"* or to be treated as lived vectors of faith or fidelity, in awaiting the result or the measure that each faithful realizes in his or her person. The plan of salvation is flexibilized, dismembered, its teleology is unmade, in particular that which solved by way of the probable and the aleatory the question of predestination or of the grace for which the faithful is responsible. There is no reified, fixed, dogmatic plan of salvation, controlled by the Church.

2.   In view of a science of religions, we have generically formalized certain Christian notions such as "Christ," "Life," and "Resurrection." This is a radical and theoretical secularization, not a historical continuity. For example, it is impossible to constitute the ultimate generic lived, that which makes

Man-in-person, using philosophical concepts or even philosophical operations. It is necessary to change style of thought and to posit it as a before-first term (but determined in-the-last-instance by its "object"). This is why what is needed here is a theory-fiction and models, a modelization. Messianity is the nothing-of-the-world, which is not a worldly nothing, not a "nihil" on whose basis something will be created in continuous and sufficient manner (ex nihilo), something that must be already given and prefigured, creation as vicious repetition. And as for "something," no radically immanent lived can derive from it. Philosophers seek a "something" only in the most remote and most anonymous regions of being, not truly beyond Being. This is to say that the messianic lived is a given, but since it is neither nothing nor something, its mode of being-given is special, it comes-without-coming, it is born-without-birth immanently, from the futural ground of itself. Is that to say that it is an object of waiting, a promise awaiting its realization? An unexpected among unexpecteds, this side of waiting itself, messianity is the futural Real rather than a realization, even if the Real under-goes immanently without returning to itself, remaining that nothing of being that it has always been.

So is it a matter of a leap in the Real, or of the last possible lived? But where would we leap from if we are already in-lived or in-coming? Perhaps we need to leap, in fact, if we are philosophers—but then we fall back into philosophy and not into the lived, from which we have never fallen. On the other hand, we never stop "refalling" in the grip of philosophy, according to an immanence that we nevertheless can never leave. How to escape from these aporias? They are only aporias for philosophical sufficiency. They surreptitiously suppose that the lived is not a simple radical idempotence, but that the Real in itself was gathered into an All at once divisible and indivisible, divided and undividable. The lived is *radically a stranger to the All, indifferent* rather than an absolute stranger, but a *Stranger-advent for* the All. Are we admitting nothing more than the Real or positing it as axiom? But if it makes possible or underdetermines the axioms, then it is itself not the object of simple axioms—it is radically anaxiomatic and anhypothetical without being that absolute of the transcendence of the Good. It would rather be a falling short of essence, more secret and futural than any falling short. To say it again and otherwise than in a Platonic manner, the Real is not anaxiomatic or anhypothetical; it is, instead, the axiom that is underdetermined unilaterally or in-the-last-instance.

3. We distinguish philo-fiction, including its religious elements, which belong to the real and not the transcendental order, from its religious modelization, which allows us to understand our usage of religion. It is religion that supplies paradigms to philosophy since the latter, considered as distinct from religions or strictly defined as transcendental mechanism, in a sense has none. Meaning that the great philosophical mutations are in fact mutations between religious paradigms.

We must fix a new primacy, a Real of a nonreligious type, a non-that is not absolute but radical. The terms used, such as Christ, Lived, and so on, are primary nonreligious terms—not immediate or absolute negations of the religious, but religious terms reduced to the state of a priori materiel and oraxiomatic fiction. On the basis of Christianity as religious paradigm, but a paradigm that is philosophized and then "transformed," here as always, into an occasion, we draw three consequences of christo-fiction.

4. The end or sacrificial consummation of transcendent religion through the affirmation of generic immanence. The two other paradigms, Greek and Judaic, conserve the transcendent real, and the Christianity of the Church and even the christo-centrism that it sometimes admits exploit it without measure, despite its end being programmed and announced by Christ, by mixing it with the Greek and Judaic. These three paradigms are linked, and they continue to be linked still in philosophy, which concatenates them with its duplicity.

5. The constitution by cloning of a Stranger-subject for whom the announced end is already realized, or immanent, is a real subject as transfinite organon or as partially dependent on the world. This subject is modelizable (and no more than modelizable) by the innumerable christologies produced by philosophy. We must thus distinguish real futurality or the specific under-going of generic Man from the messianity of the Stranger-subject, which conserves something of the transcendental.

6. This subject is the site or the ratio cognoscendi (to speak for a moment in an inadequate manner) that retroactively pronounces the ratio essendi, the consummation of the transcendent religious, or of the Jewish God, in a transfinite messianity, a futurality, or an under-going. The messianic Stranger announces retroactively that he is generic Man in-the-last-instance or that he is in-immanent-under-going.

NOTES

PREFACE: CHRISTIANITY STRIPPED BARE BY CHRIST

1. *Resumed* (*rélancé*): Laruelle's use of this word draws on the mathematical sense in which one vector can be said to resume another, that is, prolong or "relaunch" its orientation and amplitude and thereby take it further—trans.

2. *Knowledge* (*savoir*): Much of what follows depends on a distinction between *savoir* and *connaissance*, a perennial difficulty for translators. The two terms are translated here as *knowledge* (with the connotation of scholarly learning and established bodies of knowledge, and sometimes as *knowledges* [*savoirs*] when speaking of multiple such bodies) and *cognizance* (with the connotation of firsthand acquaintance or a "lived" knowledge)—trans.

3. (*Connaissance*)—trans.

1. A GENERIC REPETITION OF GNOSIS: TO DESUTURE CHRIST FROM THEOLOGY

1. *Simple ones/ Simple Souls* (*Les simples*): One of Laruelle's references in his "nontheological" works is the medieval French mystic Marguerite Porete. Her book *The Mirror of Simple Souls* is regarded as a major source for the autotheistic Heresy of the Free Spirit. Porete was burned at the stake in 1310—trans.

2. *Vectoriell* (*vectorial*, *vectoriale*): In introducing (in the French) *vectorial* as a counterpart to *vectoriel* (which means "vectorial" in the normal mathematical sense), Laruelle echoes Heidegger's distinction between *existenziell* and *existenzial* (see the introduction to *Being and Time*, section 4), usually rendered in French as *existential* and *existentiel*. The traditional English counterpart is *existentiell* and *existential*, and in line with this, *vectorial* has been translated here as *vectoriell*. The Heideggerian analogy should not be taken too literally however: The *vectoriell*, as we shall learn, is distinct from the *existentiell*: although it relates to the lived of generic man in his encounter with the world, it embodies nonstandard philosophy's insistence that the latter can be addressed through a "real (not logical) formalism." Moreover, given that the aim is not to submit man or his religion to a "positive" mathematical science, the neologism's main function is to distinguish the "ontologico-geometrical" *vectoriell* from the "psychologico-geometrical" *vectorial* and its positive sufficiency—trans.

3. *Materiel* (*Matériale*): Distinct from the "matter" posited by (philosophical) materialism, this term is adopted and transliterated into French by Laruelle from the work of Max Scheler. For Laruelle the *materiel* is the "matter" of the lived, the Husserlian *erlebnis* or the phenomenological *hyle*, which nonstandard philosophy experimentally subjects to the algebraic formalisms of quantum physics. The translation *materiel* is adopted so as to maintain it as a separate term both from (philosophical) matter and from the various "materials" that nonstandard philosophy makes use of in its procedures—trans.

7. THE TWO LAWS OF SUBSTANTIAL RELIGIOUS EXISTENCE, AND CHRIST AS MEDIATE-WITHOUT-MEDIATION

1. Paris: Harmattan, 2007.

INDEX

Abasement, 195; Ascension and, 212–14; of God, 26–28; of transcendence, 173–75

Algebra: and the lived, 103–4; in messiah-function, 50; of messianic wave, 103–16; of religions and Christ, 54–55; and superposition, 63

Alienation, 17, 190

Ambiguity, of Christianity, 164–65

Anthropic principle, x

Apophasis, 243–45

Apostles, 25–26, 40, 47, 79, 226, 255

A priori, quantum, versus transcendental, 49–50, 52

Ascension, 24, 26, 65; abasement of God and, 212–14; Cross relating to, 212; idempotence relating to, 209; messianic wave relating to, 89; nonecstatic, 209; quantum theory of, 212–13; Resurrection and, 200–1, 206–7, 212–14; Tomb relating to, 210–11; transcendence relating to, 238; *see also* Resurrection

Atheism, 2, 28, 58, 80, 128

Aufhebung, 204

Autoanalysis, circle of, 120–21

Becoming, of Christ, 149

Belief: debasing of, 80–81; as decoherence, of faith, 249–50; defining, xi–xii; faith and, xi–xii, 43, 77, 85, 97, 99–100, 241, 249–57; generic, xi, 79, 165, 180; Messiah's destruction relating to, 170–72; messianity and, 95–96, 222; science of religion and, 80–81; struggle against, 252–53; subject-in-struggle relating to, 236–37; superposition and, 97

Blood, of Christ, 222–23

Body: of Christ, 55–56, 195; humans in-body, 6, 244; and ideality, 17–18; Resurrection of, 211

Cartesian circle, 120–21
Cataphasis, 58, 243–45
Christ: ages of, 255; algebra of
religions and, 54–55; becoming
of, 149; blood of, 222–23; body of,
55–56, 195; Christo-fiction and
return of, 112–13; conversion of,
191; as founder, 123–24; generic
fusion of, 73–74; imaginary,
48–52; in-, 29–30; -in-person, 34,
47–48, 55–56, 76; interpretation
of, 130–32; invention of, 127–28;
Judaism and, 30; knowledge of,
11; as Last Instance, 200–2; on
Law, 137; logia of, 138, 153, 155–58;
material formalism and, 62–65;
meanings of, 122–23; mission of,
136–37; in non-Christian science,
122; nonsufficient science and,
19–20; -operator, 114–15; as our
contemporary, 18, 44, 75–76;
philosophy of, 29–37, 42–43;
physics of, 19–21; quantum model
of, 41–43, 55–56, 90–91; religion
relating to, 37, 76, 123–24; return
of, 51, 72, 112–13, 123; sacrifice
relating to, 218–21; state of vector,
91–93; -subject, 114–16, 122–25, 237;
as superposed, 209–10; -system,
47, 51; theo-christo-logical doublet
and invention of, 57–59; theology
under, 73–74; traditional and
Christ-subject, 122–25; unilateral
complementarity and, 124–25;
unilaterality over, 123–24; see also
Science-in-Christ; specific topics
Christianity: ambiguity of, 164–65;
Greco-, 162–64; history relating
to, 164–65; immanentization
of, 29; Laws relating to, 127–28;

non-Christian science relating to,
119–20; Paul relating to, 229–30;
Quantum theory relating to, 66;
sacrifice in, xv, 28, 29, 58–59, 66,
196, 214–15, 220; science of, 33–34
Christian science, 125–26
Christic clinamen, 195
Christic insurrection, 26–28, 84, 93
Christic matrix: composition and
construction of, 83–84; elements
of, 83–84; Jews and Greeks
relating to, 85–86; kerygma
relating to, 142–44; messianic
wave in, 100–2; messianity,
faith, and, 93–98; nontheology
criteria and, 85–87; philosophy
relating to, 85–87; PST relating
to, 85–88; unilateral dualities,
of vectoriellity, 98–100, 101;
unilaterality and messianity in,
88–90; vectoriellity in, 88, 90–91,
98–100
Christic messianity, 67–68
Christic nontheology, 57
Christo-centrism, 30; Cross relating
to, 194
Christo-fiction: Christian theology
relating to, x–xi; Christ's return
and, 112–13; procedures of,
138–39
Christology, 2, 58; microscopic, 55
Christ-science: foundation of, 179;
phases of, 100–101; quantum
orientation of, 65–68
Church: development of, 228;
faith in, 96–97, 245; on Gnosis,
4, 5; Resurrection relating to,
228–31; superposition relating
to, 170; theology, 11; universal
philosophical, 228–31

One: idempotence relating to, 171; immanence relating to, 171; in matrix, 171–72; scientific discovery relating to, 181–82; vision-in-, 181–82
One-in-person, 55–56
Oraxioms, 133–34, 156–58, 240
Original Sin, 250–52

Paganism, 28
Paul (Saint), 141, 191–92; Christianity relating to, 229–30
Phenomenological distance, 166; Cross relating to, 189–90
Phenomenology, 197–200
Philosophical circle of religion, 119–21
Philosophy, 10; of Christ, 29–37, 42–43; Christic matrix relating to, 85–87; on Cross, 64–65; errors and, 119; ethics compared to, 6–7; Hegelian, 20–21, 65–66; on humanity, 17; of in-Christ, 29–30; Judaism on, 29; Kantian, 20–21, 49–50; knowledge relating to, 13–14; nonphilosophy, 29; Principle of Sufficient Nonknowledge relating to, 13; science relating to, 16, 111–12; theology compared to, 235; wisdom compared to, 77
Physics, 33–34; of Christ, 19–20
Plato, 161
Platonism, 33–34
PMS, see Principles of Mathematical Sufficiency
Porete, Marguerite, 259n1
Positive knowledge, 8–10, 12, 129
PPS, see Principles of Philosophical Sufficiency
Principle of Sufficient Judaism (PSJ), 50

Principle of Sufficient Nonknowledge, 13
Principle of Sufficient Philosophy (PSP), 50
Principle of Sufficient Theology (PST), x, 62; Christic matrix relating to, 85–88; defining, 6; gnosis and, 6; insurrection relating to, 26
Principles of Mathematical Sufficiency (PMS), 19
Principles of Philosophical Sufficiency (PPS), 19
Proletarian science, 125–26
PSJ, see Principle of Sufficient Judaism
PSP, see Principle of Sufficient Philosophy
PST, see Principle of Sufficient Theology
Psychoanalysis, 190

Quantum a priori, 49–50
Quantum model, of Christ, 41–43
Quantum of faith, 36, 41–42, 166–67; science-in-Christ and, 76–78
Quantum-oriented order, x
Quantum-oriented theory, 66–67
Quantum physics, 34–35, 41; Christ-science relating to, 66–67; gnosis and, 75; Tao of, 77–78
Quantum preparation, 20–23
Quantum principles, 109–10
Quantum science, 117–18
Quantum-theoretical terms, 5
Quantum theory: on Ascension, 212–13; Christianity relating to, 66; of Christ's body, 55–56; of Cross, 197–200; generic, 165–67; gnostic theology in, xiii–xvi; on Original Sin, 250–52; vectoriellity relating to, 212–13

Quantum thought, 21–22
Quantum vectoriellity: Christ as, 90–91; *see also* Vectoriellity
Quantware, 21–22

Radical Christ: Christ-in-person, 34, 47–48, 55–56, 76; Christ-system and messiah-function, 47, 51; messiah-function 1, 48–52; messiah-function 2, 52–54; messianity and faith, 48; names and functions of, 46–48; science-in-Christ and, 46–54
Radical Death, 204
Radical faith, 42
Rationalization, 44
Real, 108–9
Reality condition, 92
Reason, 32–33, 36, 42–43
Reconciliation, 201
Redemption, 18
Reformed gnosis, 2–3
Religion: algebra of, 54–55; Christ relating to, 37, 76, 123–24; formal, 162; foundational religious legacies, 29; philosophical circle of, 119–21; science of, 33–34, 80–81, 165–67; substantial and generic, 161–62
Religious definition, 8
Resistance, 113–14
Resurrection, 65; abasement of God and, 212–14; Ascension and, 200–1, 206–7, 212–14; of body, 211; Christic insurrection and, 26–28; Church relating to, 228–31; cloning relating to, 210; experimental science of Cross and, 43–45; generic science relating to, 58; God's

salvation and, 223–25; humanity and, 215–17; incarnation and, 24, 25; insurrection, existence, and, 207–9; interpretation relating to, 131–32; materiality of, 209–10; mediate-without-mediation relating to, 214–15; in messiah-function, 51; messianic wave relating to, 89; messianity relating to, 203–6; as metaphysical, 51; rationalization of, 44; receiving blood of Christ, 222–23; reconciliation of, 201; Resurrected Christ, Christian inversion, and Cross, 189–92; return, science-in-Christ, and, 225–28; sacrifice and, 215–17; science of, 203–31; suspended sacrifice and, 218–22; theo-christo-logical doublet relating to, 217–18; tomb relating to, 204–5, 210–11; vectoriell relating to, 204–5; vectoriell superposition in, 206–7
Return, 225–28
Revelation: idempotence relating to, 180; scientific discovery and, 181–83
Revolutionized history, 27
Rosenzweig, Franz, 50

Sacrifice, 201; in Christianity, 214–15; Christ relating to, 218–21; of Cross, 220; of God, 215–17, 219, 221; God's death relating to, 219–20; humanity, Resurrection, and, 215–17; interpretation of, 218; Jewish persecution relating to, 224; of mediation, 149–51, 214–15, 217–18; sameness and, 219; in

Superposition, 16, 19, 22; algebra of, 63; Church relating to, 170; faith, belief, and, 97; in generic matrix, 69; of idempotence, 106; kerygma as, 142–44; messiah-function and Cross, 53–54; noncommutativity, laws, and, 140; as sacrifice, 219; scientific discovery of, 181–83; vectoriell, 63, 206–7
Suspended sacrifice, 218–22
Suture, 34, 117, 122, 222–23, 230–31

Tao, 77–78
Temporal determinism, 195–97
Temptation, 253–55
Theo-centrism, 44
Theo-christo-logical doublet: Christ's invention and, 57–59; contemporary Christ through generic science, 75–76; Cross relating to, 58–59; deconstruction of, 59; destruction of, 217–18; dualities, 59–60; foundation and generic constant, 68–73; generic fusion and theology, 73–74; generic science defined, 78–80; gnostic alliance of science, 74–75; Jews and Greeks relating to, 70–72; materiel formalism and Christ, 62–65; mediate-without-mediation relating to, 72; quantum of faith, 76–78; quantum orientation, of Christ-science, 65–68; Resurrection relating to, 217–18; science of religion and belief, 80–81; transcendence and, 60–61; as variable, 59–62; variables and inputs, 70
Theological sufficiency, *see* Principle of Sufficient Theology

Theological unintelligible faith, 37–39
Theology: basis of, 44; under Christ, 73–74; Church, 11; gnosis, science, and, 1, 15–17; metaphysics compared to, 59; philosophy compared to, 235; science-in-Christ and, 2; science of, 58; scientific deficiency of, 133–35; *see also specific theology*
Three monotheisms, 127
Tomb, 196–97; Ascension relating to, 210–11; Resurrection relating to, 204–5, 210–11
Torah, 133–34, 142–46, 213; Christ in relation to, 63, 147–48, 159; and Logos, xiv, 5, 26, 28, 35, 39, 52, 61, 80, 86, 106, 107, 127, 133, 136, 139–40, 183; superposition of, 142–46, 194, 196, 199, 213, 228, 244
Transcendence: abasement of, 173–75; Ascension relating to, 238; faith relating to, 238; immanence, sacrifice, and, 216–17, 220; immanence and, 86; theo-christo-logical doublet and, 60–61; unilateral dualities relating to, 172
Transcendental *a priori*, 49–50, 52
Transcendental illusion, 119–21
Transcendental imagination, 52–53
Truth, xi
Tunneling-effect, 238–39
Two Laws, *see* Laws

Ultimation, 234–35; conversion, universion, and, 183–84
Ultimatum, 247
Uncertainty, 97–98
Unifaciality, of real, 108–9
Unified theories, 111–12

INSURRECTIONS: CRITICAL STUDIES IN RELIGION, POLITICS, AND CULTURE

Slavoj Žižek, Clayton Crockett, Creston Davis, Jeffrey W. Robbins, Editors